New Media

Communications Technologies for the 1990s

edited by David Shorrock

London, 1988

British Library Cataloguing in Publication Data

New media: Communications Technologies for the 1990s
  1. Automated information retreival systems
  I. Shorrock, David
  001.5
  ISBN 0-86353-143-1

© Online Publications 1988

ISBN 0 86353 143 1

Typeset and printed in the United Kingdom by Henry Ling Ltd, Dorchester.

Online Publications
Blenheim House, Ash Hill Drive, Pinner, Middlesex HA5 2AE, UK
The Publishing Division of Blenheim Online Ltd, London

Acknowledgements

I would like to thank all the contributors to "New Media: Communications Technologies for the 1990s" for providing such intelligent and thought-provoking papers. In addition, I would like to express thanks to all my former colleagues at SD SCICON for their advice during the compilation of the book. Finally, special thanks to Allen Whitlington for his time and patience in the preparation of the final manuscript.

David Shorrock
London, May 1988

Contents

CD-ROM

I
Introduction

Blenheim Online is the world's leading specialist in the design, co-ordination and management of conferences and exhibitions concerned with the business implications and applications of leading edge technology. Online has held scores of international conferences that address the delivery of information by electronic means, at which some of the most authoritative specialists in the field of what is collectively termed *New Media* have presented papers.

From this library of published proceedings 26 papers have been selected from 10 recent conferences, written by some 29 authors from 7 different countries.

Each of the papers has been revised and updated by the original author to take account of the continuing developments in the technology, experience that has been gained through live applications and, with the benefit of this experience, revisions in the market forecasts. The facts and opinions are those of the individual authors and not those of the editor or the publisher.

Although new media is referred to above as the delivery of information by electronic means, other authors have explored the definition more thoroughly. Rice[1] defines new media as "those communication technologies, typically involving computer capabilities (microprocessor or mainframe), that allow or facilitate interactivity among users or between users and information.

This definition naturally leads into a discussion as to the depth of interactivity offered by various technologies. Many however, credit the Japanese with the coining the expression. The JIPDEC Report[2] offers "A good definition of new media is that they are means of communication that bring about revolutionary changes to one or more of the four processes associated with conventional media, i.e. information gathering and preparation; information processing; information transmission; and information utilisation."

However, Rice[3] reminds us that "Newness, of course, is in the eye of the cohort. At this time, we might consider 'new media' to include personal computers, videotext and teletext, interactive cable, communications satellites, office information systems and the like."

Another Japanese author, Namekawa[4] suggests that one's definition is dependent upon one's business interests: "Electronic mail, facsimile, videotext, teletext and telephone video conference are what the telepone and data communications networks call new media. Cable television, CATV, teletext, direct broadcast satellite, DBS and high-definition television are what those in the field of television broadcasting are titling new media."

In compiling this collection of papers, the editor has chosen those media that are likely to have the greatest impact on both the providers and consumers of information delivered electronically. These are:

Videotext

Audiotext

CD-ROM

Cable

Satellite

Each of these is presented in a stand-alone section, preceeded by an introductory paper by the editor, so as to set the scene for the individual papers that follow. For each of the media the papers have been selected to give a description of the base technology, the benefits compared with other media, the economics, the commercial applications and market opportunities, and, where possible, market sizes and forecasts through the 1990s.

Videotex

Global markets for videotex services

The introductory paper by David Shorrock examines the state of the videotex industry and, in particular, public access videotex systems in the UK, France, Finland, Canada and the USA.

International standards for videotex

Wolfgang Heidrich of the Deutsche Bundespost considers the development of international standards within the videotex industry and some of the problems and mechanisms for overcoming them for international videotex services. Standards initiatives being undertaken within the European Telecommunication Administrations are discussed.

Videotex in Japan: the CAPTAIN system

Toshio Terashi and Michio Sugimoto of NTT, describe one of the most advanced videotex services in the world, the Japanese CAPTAIN system. As well as discussing applications of the existing system based on analogue circuits, the Advanced CAPTAIN system based on the Integrated Services Digital Network (ISDN) is described. The predicted growth in use of the system for both business and domestic subscribers is also given.

Videotex in France: the success of Minitel

Claude Finzi of Telesystems Questel describes the French Minitel system and the explosive growth of the service, both for professional and consumer applications. The reasons for this phenomenal growth, and the likely impact it will have in France on other media such as CD-ROM, are considered.

Implementation of interactive videotex in the West German & British travel industries

Falk von Bornstaedt of the Institute of Applied Information Technology, GMD, and Margaret Bruce of the University of Manchester Institute of Science & Technology (UMIST) describe how videotex has been adopted by both the West German and British travel industries. A comparison is made between the applications and developments of the videotex systems, which reflects both the national characteristics of the industry and the way in which videotex services are provided by the PTT.

Videotex publishing: the US experience

Christopher Burns of Christopher Burns Inc., discusses both the expectations and the failures of the videotex industry in the USA. The experiences of publishers who have offered information on videotex systems, that was originally destined or taken from traditional print media, is described along with the authors views as to the reasons for the failure of such services in the marketplace.

Shopping by videotex: a revolution in retailing

Thomas Rauh of Touche Ross, San Francisco describes the characteristics and size of the electronic shopping market in the USA. He considers the economics of videotex retailing systems, the potential markets in which they may be used and suggested roles that the retailers might adopt to address these markets.

Electronic shopping for the 1990s

In this second paper, Thomas Rauh explores the videotex retailing market in greater depth. He considers the development of both public access and home-shopping systems and the possible combinations of media that they might employ, for example the combination of videotex with videodiscs.

Privacy & security on videotex systems

Hermann Maurer of the Technical University, Graz, Austria, presents a discussion on the potential threats to videotex systems (for example, loss of information, destruction and breach of privacy) and suggests mechanisms for countering them. Given the publicity attached to the exploits of the hackers who left a message in the Duke of Edinburgh's PRESTEL mailbox, and put a joke in the currency section of the Financial Times, this is a timely reminder that many of the elementary security procedures described are so often lacking from systems currently in use.

Audiotex

Audiotex: the telephone media

The introductory paper by David Shorrock gives an overview of the technology and looks to the future development of audiotex systems and their potential applications.

The audiotex industry and markets

Dr C William Reed of Link Resources Inc, and Bruce Kushnick of National Televoice describe the technologies behind audiotex and go on to discuss the market and market trends for audiotex products. The factors that have led to fragmentation of the market and its slow growth are identified.

Talking yellow pages

Leon A Ferber of Perception Technology describes how audiotex has been applied to the yellow pages service and details the benefits that have accrued to both the advertisers and the users of the service. The production process for the various types of advertiser and promoter are described.

Corporate and business applications for audiotex

Paul F Finngan of Voicemail International Inc, describes a combined voice-mail/audiotex system capable of telephoning subscribers. Two proven applications illustrate the potential of the system: the Dowphone financial information service, and the TWA crew scheduling system.

CD-ROM

Optical storage media: CD-ROM and beyond

The introductory paper by David Shorrock examines the development of CD-ROM and related optical storage techniques, CD-I, CD-V, DV-I, WORM and erasable discs. The development of standards and how this will impact the market through the 1990s is considered.

Database distribution on CD-ROM

Graham Seddon of BRS Europe gives a detailed explanation of the manufacturing processes of CDs, including the software requirements for indexing, mastering and retrieval. The economics, for an information provider, of offering a CD-ROM product as an alternative to an on-line database is considered. Some trial projects are given as demonstrations of the technology.

CD-I: future applications

Byron M Turner of European Interactive Media describes the Interactive Compact Disc, CD-I, as a synthesis of text, audio and video functions into a new publishing medium which could ultimately alter our perception of communication. The potential applications identified for CD-I are similar to those described by Thomas Rauh in his papers on videotex and electronic shopping: an example, perhaps, of new media technologies competing for the same market sectors.

CD-ROM and print: partner or competitor?

Wolfgang Benscheck of the German publishers Hoppenstedt, describes how they are using CD-ROM in conjunction with their existing print products, as part of their multi-media distribution strategy for marketing company information. The requirement to prepare information for the CD-ROM format has led to improvements in the traditional printed material.

Software publishing on CD-ROM

Eric Coates of the Digital Equipment Company discusses the feasibility as well as some of the practical difficulties of distributing computer software and associated documentation on CD-ROM. Surprisingly, it is the immense storage capacity of the disc which presents problems in establishing suitable charging mechanisms when a user only requires a subset of the disc contents.

The players in the CD industry

Anthony Chandor of Mandarin Communications Ltd describes the current state of the market and the structure and roles of the industry: the information providers, the integrators, the mastering and replication companies, and the player manufacturers. The paper includes examples of representative companies and their CD-ROM products.

The outlook for CD publishing

Haines B Gaffner of Link Resources describes the markets and their respective sizes for CD products. Some of the common myths that have grown up around CD-ROMs are critically examined and, perhaps more realistic, assessments made.

Cable

Broadband communications: a global view

In his introductory paper, David Shorrock reviews the developments in broadband networking: the various national programmes in Germany, France and the UK, the European RACE programme and the planned Global Digital Highway. The rationale behind these developments and the role of the Telecommunication Administrations and the various private operators are considered.

Broadband networking

Howard Kleyn, chairman of Oyston Cable Ltd, provides an overview of broadband telecommunications before going on to consider the handling and control of information. The importance of switching in broadband networks and their future developments and applications are also considered.

Perspectives on coding, switching & transmission in a broadband context

John Howard of Plessey Major Systems continues the discussion on broadband networking by considering the system components and their continuing development. The applications for such services, particularly video distribution, could well provide the bearer medium for videotex systems.

On-demand interactive video on British Telecom's switched star network

Gordon Kerr of British Telecom Research Laboratories describes how interactive video services can be provided over a cable system. The concept, which is currently being demonstrated on the Westminster cable system, is potentially in direct competition to the interactive CD products discussed by Byron Turner in the previous section.

The opportunities for information providers on broadband cable systems

Alan Robinson of Croydon Cable TV gives an overview of the cable television scene in the United Kingdom and suggests some opportunities the media offers to information providers. The paper also includes a commentary on some of the inhibiting factors that are slowing the development of cable systems in the UK.

The world market for broadband applications

David Rumble of PA gives a global view of market for broadband services. The paper not only addresses which applications are likely to succeed over the next few years, but also: who will pay for them?

Satellite

Satellites and small dishes: business or pleasure?

In this introduction to the final section, David Shorrock describes how the economics of satellite communications have changed with the development of small dishes both for television reception and for business data transmission and reception. The development of what are referred to as 'very small aperture terminals', (VSATs), or 'microterminals' is described along with an overview of the technology, the benefits and some of the applications. A novel application of the European experimental satellite Olympus is also described.

Data broadcasting by satellite on Europe

Lionel Fleury of Polycom describes the first European satellite data broadcast service, which commenced commercial service in France in March 1987. The company, Polycom, is able to overcome many of the restrictions governing satellite communications because it is (indirectly) a subsidiary of the French PTT, and hence can exploit the PTT's privileges.

Towards the intelligent satellite

Professor Barry Evans of the University of Surrey describes the current state-of-the-art in communication satellites and the developments that are leading to the 'intelligent' satellites of the 1990s and beyond. By placing intelligence on-board the satellite, the economics of information delivery by satellite could be transformed so that it becomes the cheapest and most pervasive of the new media.

16 Channels & medium power: the logical way ahead

Marcus Bicknell of Société Européenne des Satellites describes ASTRA, Europe's first private satellite, scheduled for launch towards the end of 1988. He describes the advantages that ASTRA offers compared with both the existing telecommunication satellites used for television distribution, and the high powered national direct broadcast satellite (DBS) systems that are planned. The paper contains detailed statistics on the European markets.

DBS—understanding the risks and opportunities

Jon Chaplin of the European Space Agency discusses the risks and opportunities facing the high-power DBS systems proposed for Europe. These risks and opportunities affect the system operators, the programme providers, the receiving equipment manufacturers, and ultimately, the home consumers who will directly or indirectly govern the success or failure of these projects.

Summary

From the papers selected it can be seen that there are no clear boundaries between the various media. Many of the applications described can be satisfied by one or more of the media, either alone or in combination. Their use is dependent upon various factors including: the nature of the information to be disseminated; the timeliness of that information; the number of consumers, their degree of sophistication and/or familiarity with the media, their freedom and frequency of access to terminal equipment; the economics of the distribution process; the competitive threat posed by alternative media; and finally, a subject which has not been touched on here, the legal and regulatory environment applicable either to the information or the use of the media.

The selection of a given medium is, therefore, a complex choice. It is hoped that the papers in the following sections give as complete a discussion of the relevant factors as current knowledge will permit or, at least, as is possible in a single volume. From this discussion the reader can make his own choice as to the suitability of a given medium to a particular application.

2
Global markets for videotex services
David Shorrock

What is videotex?

Videotex has been defined[1] as a "system for the widespread dissemination of textual and graphic information by wholly electronic means for display on low-cost terminals (often suitably equipped television receivers) under the selective control of the recipient, using control procedures easily understood by untrained users". This definition encompasses what was considered to be the two primary advantages of videotex: that it is based on a low cost and readily available display technology, the television receiver, and that it is a medium that can be used by the casual user with little or no training.

The Comité Consultatif International Télégraphique et Téléphonique (CCITT) describe videotex as a generic term for two specific types of service: broadcast or one-way videotex, often called teletext and interactive videotex, often referred to as videotext.

The market

First launched in the United Kingdom onto the domestic market, public service videotex all but failed until it was effectively relaunched for the business community and, in particular, the travel and insurance industries. Some limited domestic application has been found in home banking. In contrast, videotex in France has been successfully launched onto the domestic market, on the back of the electronic directory enquiry service, but now generating over 70 per cent of revenues from video games and other personal services. In the US, after false starts not unlike the British experience, videotex is now seen as heralding a revolution in retailing, electronic shopping, perhaps confirming the phrase of the American businesswoman 'born to shop'.

In corporate applications, the US and European markets have also differed. More than 1,000 companies in Europe are now using videotex as a means of providing low cost, easy to use, access to the corporate computing facilities. In contrast, the American experience has been of a slower take-up. One of the reasons suggested for this, is that videotex has been identified with the consumer systems and their failures. Another reason is perhaps the high level of corporate penetration achieved by the personal computer in American companies and the consequent computer literacy of the users and therefore the lesser need for simpler systems. However, as discussed elsewhere in this section, corporate videotex applications are now beginning to grow in the US.

Videotex in the UK

The perceived advantages, low cost TV sets and ease of use, shaped the first applications of the media. The first commercial service was established by the British Post Office (now British Telecom, BT) in 1979, under the title 'Prestel'. Prestel was seen as a way of generating additional revenues from under-utilised domestic telephone lines, by using them to deliver information electronically to households. The idea was that information providers would feed central databases, owned and operated by BT, and that users would, having bought a modified TV or adapter, be able to call up 'pages' of information for a small fee. The age of the electronic newspaper was born. However, the hundreds of thousands of domestic subscribers predicted in the original business plan never materialised. The Prestel TV set (or adaptor) cost hundreds of pounds and was incompatible with the teletext service offered by the broadcasters, who also provided their information free of charge, whereas for Prestel every page had to be paid for, in addition to the telephone call charge. The domestic subscriber realised what excellent value newsprint really is.

So Prestel grew slowly as niche markets developed in specialist information, of interest to business rather than domestic users. Today it is estimated that there are some 76,000 Prestel subscriber terminals in the United Kingdom, of which 61 per cent are business users. The service has 1,200 information providers with some ten specialised services such as the information providers for the travel and insurance industries. Electronic yellow pages have been recently introduced, but as yet there is no electronic telephone directory enquiry service. This is in stark contrast to France.

The French Teletel Service

The French state telecomms utility, the Direction Générale des Télécommunications (DGT), has created the Teletel programme to encourage the coherent development of on-line data centres and videotex services.

The first pilot Teletel system was trialled in 1981 at Vélizy, a suburb of Paris, when 2,500 'Minitel' terminals were distributed free of charge. The original application of Minitel was for accessing the videotex telephone directory service, in place of the printed paper directory. The DGT then decided to connect Teletel to Transpac, the national packet switched network. This allowed subscribers throughout France to use other services, regardless of distance. At the same time, information providers found that a new market had been created by the introduction of the 'Kiosque' or news-stand billing system, whereby individual users can anonomously access the information provider's database without having to be a subscriber to the database. The information provider bills the user by means of the telephone billing system, with the payment distributed according to the traffic statistics, hence the more popular the service, the greater the revenue.

Videotex was introduced in France, as was the intention in the UK, to prolong the growth of the public switched telephone network (PSTN) once the demand for basic telephone service had been satisfied. It was also considered that the paper telephone directories were becoming too voluminous, as well as increasingly difficult and costly to update and distribute, hence the Electronic Directory Service. Minitel terminals are now offered free of charge to telephone subscribers throughout France, in place of the paper directory.

There are now in excess of 3 million Minitels installed, with the numbers growing daily. The range of services available includes continuously updated news bulletins, video games, price comparisons between supermarkets, hotel, airline and rail reservations and a whole range of services that are best described as 'of a sexually explicit nature'.

Finland and France

Finland has operated a public service videotex system, VDX-100 since 1984. The service was, until recently tri-lingual: Finnish, English and Swedish, but now a fourth language is to be added with the Finnish PTT announcing interconnection with the French Teletel service.

VDX-100 users will have access to all the kiosk services by selecting Teletel France on their screen menu. Interworking between the two systems has been eased by the fact that they are similar in architecture and that they both use the X.25 packet switching standard provided by the Finnish Datapac and French Transpac services.

The Director General of the Finnish PTT has been quoted as saying that "in ten years, there will be a difference in the economic health of countries which make a serious commitment to videotex for consumers and business. That is one reason why Finland is offering the Teletel connection".[2] The initial beneficiaries of the interconnect are expected to be companies having business with France, such as importers and exporters, banks, transport and the travel industry.

Canada adopts the Minitel

Bell Canada are planning to launch Alex, a public videotex service. Alex will use the French Teletel service as a model, acting as a gateway to services offered by other information providers, although Bell do intend to offer an electronic directory enquiry service themselves. The service, which is scheduled to commence in September 1988, will initially serve some 10,000 terminals within the Montreal area. In contrast to the French approach of giving the terminals free to existing telephone subscribers, Bell Canada intend to charge in the region of $15 (Canadian) per month for the terminals.

Videotex in the USA

Videotex in the USA has been characterised by some highly publicised failures in the home information market. Late 1985 and early 1986 saw the failures of Knight Ridder Newspapers, Viewtron, Times-Mirror's Gateway, and Centel Corporation's Keycom service. Yet despite this, some major new videotex projects are underway which will attempt to re-enter the information market as well as introducing home shopping. Critics of the failures blamed the high charges, so these new services are aiming to be low cost, with a joining fee of $30 and monthly subscriptions of between $12 and $15.

Prodigy: videotex for the home

The most ambitious of the new services is Prodigy, under development by Trintex, a four-year-old joint venture between IBM and Sears Roebuck. Prodigy is scheduled for launch in March–April 1988, either in Hartford, Conneticut, or Westchester County, New York State. The service will offer news, sport and weather information, TV and film listings and reviews, electronic mail and a brokerage service. Also under discussion are gateways to external databases such as Nexis and Dialog. The pricing of Prodigy has been described as being the equivalent of a premium newspaper subscription or cable television fee. In order to get within the $12 to $15 per month window, the service will rely on advertising revenue.

Trintex are estimated to have some 600 staff working on the project and to have invested more than $250 million. The system, which will be based on a distributed network of some 1,000 IMB Series 1 minicomputers when it is eventually offered as a national service, is predicted to show a profit within seven years of commencement of the service. What must pose the greatest uncertainty to the success of the project however, is the requirement for users to have a microcomputer and suitable modem to be able to access the service.

Telaction: electronic shopping

Telaction is an innovative attempt to provide low cost access to a videotex system which has been trialled by 125,000 households in the Chicago area. The service, which offers electronic shopping from home, is based on a designated television channel of a cable TV network. The user selects the TV channel with the Telaction logo, then dials a local telephone number using a touch-tone telephone. Once the call is established, the user can call up pictures of the various items by pressing the phone buttons. The service is effectively an electronic shopping mall where users can browse at will, holding the goods they have selected in an electronic shopping trolley, before sorting through them at the end to decide what to buy. The user must have a personal identification number (PIN) to make a purchase.

Telaction is a wholly-owned subsidiary of JC Penney, with equipment supplied by Cableshare of Canada. In particular, Cableshare supply 'framegrabbers' which receive the pictures and sound from a video storage system and send them down the cable channel which is arranged as a party-line of between 15 to 40 households. The disadvantage of this system is that all the households on the party-line can see the items that are being viewed by the purchaser, but despite this, it is claimed that individual access is still available for 90 per cent of the time and that users are spending an average of 20 to 25 minutes per day on the system.

The Chicago system currently has 40 retailers who each pay a rental fee and a percentage of their sales to Telaction. With the trial now considered to be a success, Telaction are looking to expand into other metropolitan areas in the USA, as well as considering overseas expansion.

IBM: providing the backbone for international videotex

IBM operate a global teleprocessing network, International Business Services (IBS), that serves 52 countries from four major computing centres: Warwick in the UK, Zoetermeer in the Netherlands, Tampa in the US and Tokyo in Japan. The IBS network is being used by a number of videotex service providers. Examples include international goods transporters, a US bank using the network for validating travellers cheques, and the Swedish company ESAB, using the network for inventory management in 40 factories across 25 countries.

However, the international transport services are perhaps benefitting most from running videotex across IBS. The French LAMI professional transportation service is available on Teletel but, without a similar penetration of public videotex services in other countries, would not be able to offer an international, or even European service. In order to overcome this, Transpotel International are using IBS to connect the various national videotex networks offered by the PTTs to local IBM databases. These are all serviced by the computers at Zoetermeer. A user enters the Transpotel service through a videotex terminal, which then provides access to the central database via IBS.

The Transpotel service, which is in effect an international value added network, has over 1,100 users, of which some five per cent are manufacturers and the remainder, forwarding agents and shipping companies.

Summary

The future of videotex remains uncertain. In France, public access services have undoubtedly been a success. How much this is due to the French fascination with technology is unclear. What is certain is that, by giving the terminals away, a large installed base is assured and it is the size of the installed base that governs whether information providers will find the market attractive and offer services. Whether the Canadian Alex service will be successful, with users having to pay a monthly subscription, as opposed to the French pay-per-use approach, only time will tell. Perhaps it was an act of faith, rather than any reasoned commercial judgement on behalf of the French DGT, that led to the development and free issue of the Minitel.

The US Prodigy system is an enigma. It is a considerable act of faith to invest $250 million into a system in which the payoff is seven years distant. What is more surprising however, is the requirement for the users to have a personal computer and special modem to access the service. Although PC-based videotex terminals offer greater functionality, local storage and processing, their capital cost is still in the region of $1,000. Market research amongst PC owners in the US shows that many owners already have slow-speed modems for accessing local bulletin boards, software exchange groups and on-line databases such as The Source and Compuserve: services which are already offering many of the facilities that Prodigy will offer.

Perhaps the greatest area for future growth of videotex will be in combination with other delivery media: cable, satellite or CD-ROM. The Chicago Telaction service is a good example of this. It also illustrates an innovative use of the original definition of videotex: ". . . for display on low-cost terminals . . . using control procedures easily understood by untrained users".

3
International standards for videotex
Wolfgang Heidrich

Introduction

The international videotex service will surely gain great importance in the years ahead. This paper deals with its development at the Deutsche Bundespost (Federal German Post Office) and the significance of the associated protocols and regulations.

Imagine that you had just invented the motor car. It runs quite well, although it might be somewhat more comfortable. While still working on its improvement, you have the chance to sell a number of vehicles. Some buyers have recognised the trailblazing concept of your invention and develop it further themselves. One customer, for example, has found out that the car can be used, in addition to passenger transportation, for drawing loads. In other words, he has discovered the principle of the trailer and the trailer coupling. This brilliant idea makes your invention still more valuable. After modifying the construction plans, you are going to implement it in your future vehicles thereby ensuring even better sales. New ideas are born by you and your customers, put into practice and used to enhance or alter your invention even more. An end to innovations is neither in sight nor to be expected.

This car and trailer coupling analogy serves well to demonstrate the evolution of videotex and the applicable national gateway protocols. In the course of this paper I will repeatedly return to this analogy, as it appears to be well suited, in my opinion, to give even newcomers to the field of videotex a clear picture of its development. However, it is not too difficult to reconstruct this evolution from the configuration of the telephone, television receiver and data base to the systems currently in use. The history of videotex has been, and still is, characterised by both splendid ideas and by grave mistakes, by both missionaries and itinerant preachers.

This introduction serves as an indication that videotex once again is in the process of undergoing a decisive change. The second trailer coupling, which, for technical and operational reasons, cannot be identical with the first one, is just under development and will be installed in the years to come. It is the international videotex interworking protocol (termed VI-Protocol), intended to enable a border-crossing videotex service. The immense value of the new coupling is discussed later. Suffice it to say here how important it will be to make this coupling available shortly and to endow it, for economic reason, with the features of its forerunner, so that future videotex systems need not be equipped with two couplings.

Let me now give you a more detailed description of the novel coupling under development, which consists of a new interworking protocol and, as a rule, of gateway processors. I will do this in three sections. First, information from the development workshop and on the preparations necessary to achieve maturity of series production; then the differences between international and national protocols and the additional requirements to be satisfied; finally, the purpose and service aspects of an international videotex system.

Development

A while ago, I defined the first videotex coupling. It corresponds to the computer interworking protocol developed by my colleagues at the Fernmeldetechnisches Zentralamt (Telecommunications Engineering Centre) at Darmstadt in 1979/1980, which was demonstrated to the public for the first time in London on the occasion of a live transmission to the mail-order house 'Quelle' in Nuremberg. This protocol is known today under the name 'Prestel-Gateway'. Prestel acquired its marketing rights later.

The protocol in question is asymmetrical, that is, suitable only for dialogue operation between a videotex service and an external computer, but not vice versa. It rapidly found wide application because at that time—with two exceptions—all Europe-based trial and commercial videotext systems came from a single manufacturer and made use of the same presentation standard. The normally rather long standardisation phase could be dropped: reality set a *de facto* standard.

Encouraged by the favourable experience gathered from the national computer interworking system, design engineers from three countries decided to extend the protocol in such a manner that not only are national computers accessible via the videotex service, but they can also communicate with the videotex systems abroad. For this reason, data links were established—as early as 1982—between the test systems of the Dutch PTT (Viditel), the Deutsche Bundespost (Bildschirmtext) and British Telecom (Prestel). Unfortunately, however, this second pioneering development was abruptly interrupted when the Deutsche Bundespost resolved, in September 1983, to open the general Bildschirmtext service based on the new CEPT standard. Since that task required the joint efforts of all the staff available and the conversion from Prestel to CEPT was accompanied by a change of systems, the international project was postponed. Only after a three-year respite was work resumed on an international scale within CEPT. It is now drawing to a close. The new VI-Protocol has been completed now, and the implementation manual is currently being written. The remaining countries, in addition to numerous videotex users including the Commission of the European Community, are observing these activities with keen interest.

Protocols

Let us leave the development workshop now and move on to the product itself. What is so remarkable about this protocol? As mentioned earlier, it must be able to meet the performance characteristics of the former gateway protocol—that is, to allow a dialogue between a videotex service and the external computers connected to it, where the latter may stem from any supplier and have any configuration whatsoever. This feature of the VI-Protocol cannot be stressed frequently enough, since nothing comparable existed before. On top of this, dialogue operation must be possible in both directions: from the service in country A to that in country B, and vice versa.

System designers were confronted with a further problem. In contrast to the situation prevailing in 1982, the evolution of the national services have drifted apart in the absence of adequate international recommendations. Whereas the service profiles and even the subscriber commands were largely identical until 1982, at least in the Prestel-oriented countries, they developed afterwards in slightly different directions, depending on the national requirements to be fulfilled and on the whim of the designers. Regrettably, even identical new facilities were realised in a different fashion. These discrepancies have to be partially bridged by the protocol, at considerable cost. The fact that this cannot always be fully achieved will be discussed later on.

Another adaptation that has to be accomplished by the gateway processors rather than the VI-Protocol concerns the presentation standard. In 1981, the CEPT member countries agreed on such a common standard in general, and—a few years later—in detail. This new CEPT standard unified the then separate Antiope and Prestel standards and, at the same time, enlarged the scope of facilities (among others) in the fields of alphanumeric/graphic characters and colour display. It permits all Latin-based languages to be represented; other scripts and characters can be selected by means of downloading. The same mechanism also enables high-resolution graphics to be displayed and the number of colours has been increased from 8 to 4096. The unification of the two original standards and the substantial enhancement of the range of facilities, including the geometric and photographic options, have turned the CEPT standard into a most efficient presentation tool which is gradually replacing the old standards. It goes without saying that any new services developed will make use of this standard. When the Deutsche Bundespost converted its technical system, it did not hesitate to adopt the new CEPT standard immediately.

But back once more to the protocols. The essential distinction between protocols 1 and 2 is that now, in international traffic, administrative information has to be exchanged between the individual services/administrations. The information providers want to collect money for services obtained from abroad, the administrations wish to be paid for the provision of circuits and technical equipment and the subscribers to the messaging service, for instance, would like to know in advance whether the recipient of a message really exists and has opened his electronic mailbox. The fulfilment of these requirements accounts for the greater part of the new protocol. I now intend to give a more detailed description of some problem areas that have to be coped with. However, videotex providers and subscribers should also be familiar with these difficulties and recognise that the demand for new facilities will not only concern the national services in the future, but frequently will have international repercussions. This is why, in the event of modification requests, the profiles of other services should be taken into consideration as well.

Services

International registration, transfer and accounting of the videotex providers' remuneration constitute the main problems to be solved. Whereas in the first systems payment was effected exclusively on a per-page-basis, time- and even volume-dependent charges were raised gradually. As long as all services are provided with the full range of billing mechanisms, international accounting does not pose serious difficulties but, in the absence of one or the other principle, grave conversion problems could ensue. A further dilemma arises from the frame and system field formats. There are services with up to 9-digit frame numbers, and others which use as high as 16-digit numbers, while the French system operates with no page numbers at all, but with mnemonic code names in some cases. Also, the length of the subscriber number and the size of the data fields destined for the name and address differ. Here, a few fundamental modifications and additions to the existing system will be required before an international service can go into operation. A quite different matter which poses further problems are the subscriber commands used for information retrieval and in the dialogue mode. The services are configured so differently that a user-friendly conversion scheme is not feasible.

Finally, I would like to draw direct attention to three problems that have not been solved so far. Though of a national nature, they are exclusively connected with the international service. The difficulties concern value-added tax, customs duties and the conversion of currencies; where the latter is concerned, administrations can easily fall prey to currency speculators. The utmost alertness is advisable in this respect. It is to be hoped that the installation of the new 'coupling' for videotex and the necessary tests will run smoothly and end successfully. Unfortunately, as mentioned earlier on several occasions, all existing systems have to be complemented and modified for insertion into the new coupling—just as is the case when a car has to be fitted subsequently with a trailer coupling. Wires must be pulled in for the lighting system and direction indicators, an interchange of the indicator relay and mounting of a control lamp are necessary.

Standards

A paper on the international videotex service would be incomplete if the significant topic of standardisation were skipped. I have already devoted myself, for a number of years, to this field of activity which takes up an appreciable part of my working life. After some time, one gets used to considering this problem not only from the technical but also from a rather philosophical point of view. Technological approaches are often discovered in a relatively short period of time. However, since the adoption of the resultant recommendations does not always depend on rational decisions, satisfactory solutions are frequently delayed until they have become obsolete and useless. If standards are not agreed upon quickly, outward pressure may lead to the emergence of incompatible national services whose international unification, as a rule, is likely to become nothing but patchwork, with intolerable quality losses. Unluckily, there are quite a number of sad examples for this, not only in the world of communications but also in the railways sector—to stick to couplings.

Standardisation should not be supported at any price, but public world-wide telecommunication traffic cannot be handled properly without international norms. So, if private enterprises or even administrations think it necessary to launch incompatible products, they should at least refrain from obstructing or preventing international harmonisation activities. It will, no doubt, still take a number of years to remove the divergencies that have accumulated in the field of videotex. In the interests of a rapid growth of services and traffic, all forces should be mobilised so as to arrive at an early standardisation.

Forecasts

Assume that system implementation and expansion have been completed and that the international videotex service has successfully been implemented. Who will be its first customers? True to the slogan 'money makes the world go round', the initial users are supposed to come from the commercial sector. Big international and national companies with a high export share have been eagerly waiting for the opening of an international videotex service, because it will offer them a low-cost and simple means of exchanging information with their foreign representations, subsidiaries, branch offices and traders, unless they have already been using the existing data networks. But, even assuming previous use of existing data networks, videotex will be of interest to many people, since it allows a large number of work stations to be connected cost-effectively to a central processor. Videotex as a communications path for trade and industry is an application which its inventors would certainly not have envisaged.

Enumerating all the business applications which could take advantage of videotex should ideally, be left to market researchers and marketing experts. I shall restrict myself here to giving you some typical examples: transport, tourism, the car industry and, surely, science as well. Let me consider the latter in some more detail since new aspects are practicable thanks to videotex. Data bases are becoming increasingly accessible via the national videotex services. In this respect, videotex is a key that opens many doors. What in the past was reserved for a privileged few—be it for cost reasons or because of insufficient practical training—is now becoming more accessible. The nearly immeasurable bulk of knowledge from all disciplines, collected in data banks, can be called up by everybody via easy-to-handle terminals, with simple search words and at moderate cost. Why, then, should this rapid development, which currently takes place at the national level, not be possible internationally, too? To my mind, this is one of the great chances of an international videotex service that ought not to be missed. Businesses will make use of the existing data networks if it makes commercial sense and no cheaper alternative means of communication is at hand. For private persons or minor research fields this will normally be too costly, but videotex opens up a new dimension. Any administration which, for admittedly sound reasons, does not introduce the international coupling, will withhold an important service from its customers. For the sake of completeness, it should still be mentioned that in exceptional cases external computers can also be directly connected to videotex services abroad. Whether this is economical will depend, among other things, on the amount of traffic to be handled.

Speaking about prospective users, we must, of course, not forget the many people living abroad, for whom videotex—in addition to the telephone and radio or TV—may become an essential part of their lives. International videotex will enable them to use easily the videotex service of their native country with all the facilities it offers: fast reception of the latest news in their mother tongue, transmission of messages, placing orders and keeping contact in writing with their community or club. The international videotex service opens up not only an economic but also a sociopolitical perspective which should not be underestimated.

Conclusion

It stands to reason that the acceptance of the service will strongly depend on its costs and the quality of the man-machine interface. Let us return again to the example of the trailer coupling: I would surely buy myself such a coupling if it fulfills what I expect from it—namely, that it draws my caravan in the holiday season. Its price must, of course, be reasonable. Of equal importance is its performance. If the coupling cannot bear the weight of my trailer, then I will not buy it—if the profile of the international videotex service does not sufficiently support the application which it is intended for, I will not use the service. With such an argument in mind, stringent requirements have to be placed on the new protocol and its implementation in the national services.

4
Videotex in Japan: the CAPTAIN system

Michio Sugimoto & Toshio Terashi

Introduction

The Character and Pattern Telephone Access Information Network (CAPTAIN) system, the first and the most standard videotex system in Japan, has been providing the commercial service since November 30 1984. The service is now available in all major cities throughout Japan.

The Information Providers (IPs) have been providing various services, using numerous display functions, based on CAPTAIN PLPS, and many application functions of the CAPTAIN system. For the purpose of popularising the CAPTAIN system, NTT has incorporated user's and IP's suggestions into the system enhancement plans.

On the other hand, the research and development of the digital videotex communication network, the next generation CAPTAIN system, is being promoted strongly. In a few years, the system will be introduced to the CAPTAIN service. Moreover, apart from the above-mentioned public-type system, the small-sized private CAPTAIN, has been introduced in various business fields; digitisation technology is also being adopted. These advanced systems are expected to play an important part in the highly advanced information society.

CAPTAIN System outline

The CAPTAIN system is composed of three principle parts—the videotex communication network, the CAPTAIN information centre and other external information centres, and user terminals. An outline of the CAPTAIN system configuration is shown in Figure 4.1.

The videotex communication network has been designed as a public network. It is composed of the videotex communication processing unit (VCP), the public telephone network (PSTN) and transmission lines connecting these equipments. VCP plays a central role in the videotex communication network. It possesses various communication processing functions such as protocol conversion between the information centres and user terminals, and code/pattern conversion according to the rank of user terminals.

The information centres are classified into two categories. One is the shared CAPTAIN information centre developed and supplied by NTT. The other is the external information centre owned by the IP. The CAPTAIN information centre has large capacity files. IPs without their own computer centres can provide a wide variety of videotex services economically, using files of the CAPTAIN information centre.

Moreover, the CAPTAIN information centre has various application functions, especially the retrieval, input and update of graphic information frames. In addition it can provide the order-entry service (such as mail order and reservation) the closed user group service, a simple calculation service and the gateway service to handle an external information centre.

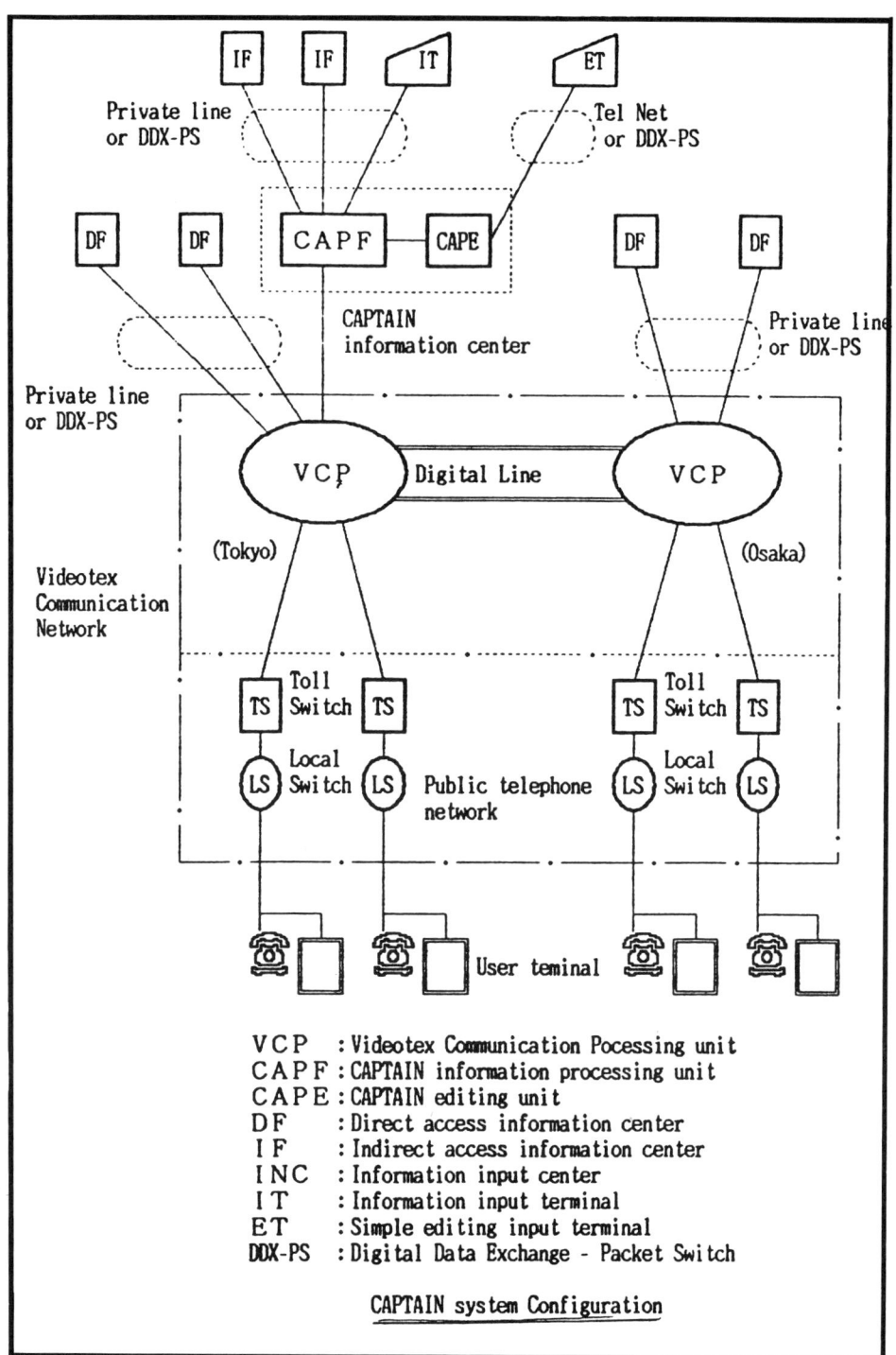

Figure 4.1

There are three different kinds of external information centres, determined according to the ways of connection to the videotex communication network or to the CAPTAIN information centre (Figure 4.1): Direct Access Information Center (DF); Indirect Access Information Center (IF); Information Input Center (INC).

Information input centre (INC)

An input information centre connects to the CAPF in the same way as the input terminal (IT). That is, the input information centre inputs and updates information frames stored in the database of the CAPF automatically. The input information centre has no direct conversation with the user terminal. This type of service is suitable when frames are updated in large quantity and frequently, like the stock dealing and quotation service.

Indirect access information centre (IF)

An indirect information centre is connected indirectly to the public CAPTAIN network via the CAPF. Being connected to the CAPF, the indirect information centre can utilise various CAPF functions such as conversation with user terminals, order entry and the closed user group (CUG) function.

Direct access information centre (DF)

A direct information centre connects directly to the public videotex network. The direct information centre itself can interact with user terminals; information retrieval, order entry and computer processing are possible if the direct information centre supports them. The burden of development is heavier than for the indirect information centre and the input centre because IPs must implement all the necessary work by themselves.

CAPTAIN PLPS

The CAPTAIN system can possess various advanced functions, based on CAPTAIN PLPS. CAPTAIN PLPS is Data Syntax I of CCITT T.101 and is based on a hybrid coding structure, using both a pattern mode for the transmission of photographic images and a coding mode for the joint transmission of characters and graphics. It can display all three major types of graphics for videotex: the geometric graphic, the mosaic graphic and the photo graphic. The standard resolution is 204 dots vertically by 248 dots horizontally. High resolution is 408 dots vertically by 496 dots horizontally. With a high resolution display, CAPTAIN PLPS can present 480 Chinese characters or 1,920 smaller characters, and four display frames of standard resolution at the same time. CAPTAIN PLPS can also provide simple animated moving, melody and tele-software functions.

Present Condition of CAPTAIN Services

As of the end of 1987 there were more than 600 IPs, providing information and services via the CAPTAIN information centre in which about 200,000 display frames are stored. The number of external information centres corresponding to DF, IF and INC type are 62, 15 and 16 respectively. The IP classification by industry is shown in Figure 4.2.

Increasing trends of CAPTAIN user terminals, and of IPs and display frames, are shown in Figures 4.3 and 4.4 respectively. The rate of user terminals settled in offices is 65 per cent and that settled in homes is 35 per cent.

Monthly access is about 16 million display frames. The three most accessed information categories are games, electronic mail, and stock quotations. The usage rate of the CAPTAIN information centre is more than 19,000 calls per day on weekdays. Average holding time and average number of frames per call are about six minutes and about 40 frames, respectively.

Classification of Industry

Category	IP of CAPTAIN information center	IP of DF center
Construction , Real Estate	1 2 IP	1 IP
Manufacture (Food / Electric, Precision Machine)	6 6 IP	1 2 IP
Publisher, Printing	4 1 IP	2 IP
Commerce (Department Store, Supermarket / Trading Company)	8 5 IP	2 IP
Financier (Bank, Financial Bank, Insurance / Stock Company, Credit Card Compan)	1 2 0 IP	9 IP
Transportation	1 0 IP	2 IP
Communication (Broadcasting Company, Newspaper / News Agency , Telecommunication)	4 7 IP	1 IP
Electric, Gas Company	4 IP	—
Information	8 2 IP	1 6 IP
Advertising agency	2 4 IP	1 IP
Travel Agency	9 IP	—
Service	6 1 IP	—
School , Hospital	1 2 IP	—
Public corporation	6 3 IP	1 IP
Organization	2 7 IP	2 IP
Other	2 2 IP	—
TOTAL	6 8 5 IP	4 9 IP

Figure 4.2

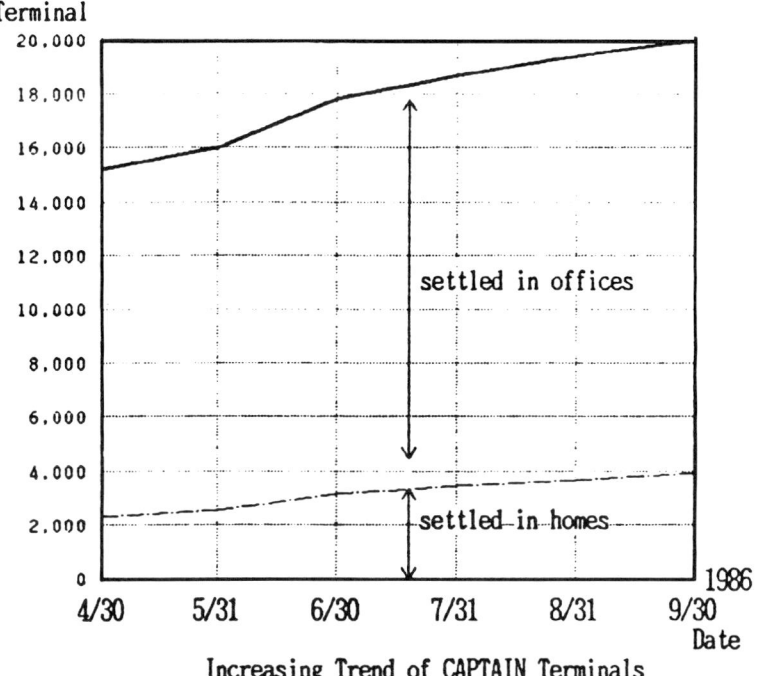

Figure 4.3

Increasing Trend of CAPTAIN Terminals

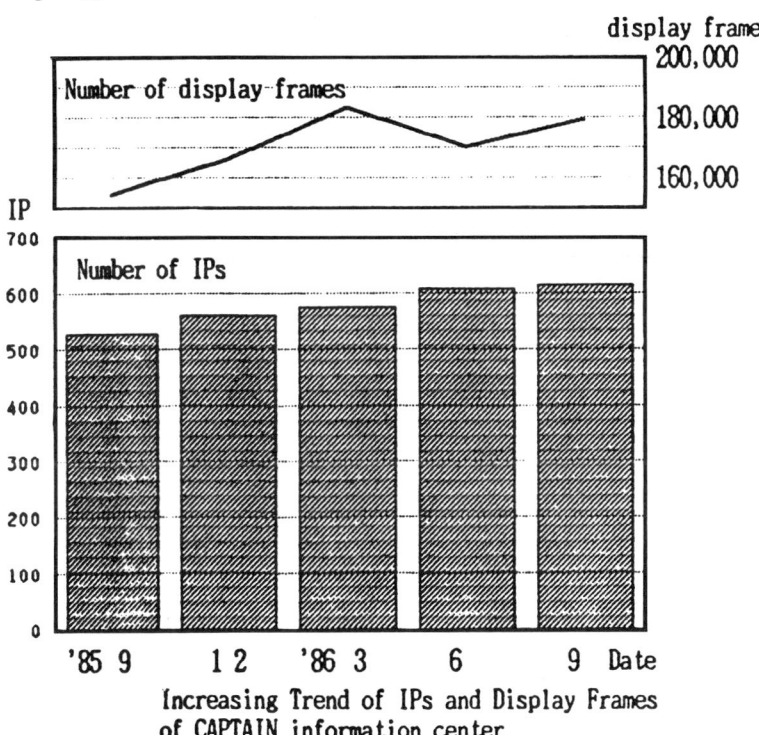

Increasing Trend of IPs and Display Frames
of CAPTAIN information center

Figure 4.4

There are many application services, provided by CAPTAIN IPs, which have good reputations among users. The following are just a few typical and interesting examples of these services.

Used-car auction service

Online used-car information service is available in the CAPTAIN system. There are about 400 dealers using the system, and more than 8,000 records of used-cars are stored in the system. CAPTAIN users can get the used-car data by using reference terms such as model, price, kinds of usage and dealer etc. As the used-car stock information has to be updated frequently, this system makes it possible to do this by using any user terminal in the dealer's office, thereby providing the most accurate used-car information. Moreover, car auctions are held on the CAPTAIN system allowing dealers to take part without travelling to the auction site.

Medical information service for doctors and pharmacists

Useful medical databases have been built-up in medical service organisations and companies. Some of these databases are connected to the CAPTAIN system, and provide various medical information services to doctors and pharmacists.

Travel information and hotel reservation service

Travel information and hotel reservation service, provided to branch offices of travel agencies, are available on the CAPTAIN system.

Education information service

Entrance exams information, a guide to schools of one's choice and other educational information are available in the CAPTAIN system.

Stocks dealing and stocks quotation service

Stocks dealing and stocks quotation services are provided by some stockbroking companies.

Electronic encyclopaedia service

The encyclopaedia database is connected to the CAPTAIN system and provides an electronic encyclopaedia service.

Digital CAPTAIN network

NTT has been pushing ahead with the establishment of Japan's integrated service digital network (ISDN) in order to provide the advanced telecommunications functions that will be needed in tomorrow's information-intensive society. Because the videotex communication network depends on the PSTN it is necessary that the videotex communication network is digitised at the same time as the PSTN.

Digital CAPTAIN, the next generation CAPTAIN system, will make it possible to send high quality audio signals and full colour photographs. The digital CAPTAIN system will be totally based on the ISDN digital network technology on which transmission speeds of 64 K bits/s will be employed between the network and the terminals.

An experimental digital CAPTAIN system, 64 K bits/s service, has been implemented since September 1984, providing an experimental service in the district of Musashino city and Mitaka city, suburban cities of Tokyo as a part of NTT's INS model system. These digital services were also provided in the Tsukuba scientific international exposition, 1985. The commercial services for digital CAPTAIN system are planned to start in 1989.

An outline of the future CAPTAIN system configuration is shown in Figure 4.5. This system is composed of information centres, user terminals (both analogue and digital) and the digital CAPTAIN network. The digital CAPTAIN network is composed of digital VCP (DVCP), the existing packet-switched data network, common channel signalling (CCS) network and the digital network.

The digital network, which includes the digital switched network and existing PSTN, has the function of converging calls from analogue and digital terminals. The communication processing Gateway Switch (GS) has functions common to all communication processing systems, such as subscriber information management, charging, traffic control and call distribution. Control signals between the GS and DVCP are transmitted through the CCS network. The digital subscriber loop provides a 64Kbits/s information channel.

The existing information centre protocol and analogue terminal protocol will also be supported in the future system. The digital terminal protocol will be able to provide more advanced and sophisticated services. The presentation level protocol has new facilities for full colour pictures and voice in addition to the existing facilities.

Information centres will be able to provide various enhanced new services to end-users using enhanced network capabilities and facilities in the future system. A full colour picture compressed by the Adaptive Block Truncation Coding (ABTC) method is provided to digital terminals. For example, a high-resolution full colour picture-composed of 408 dots vertically and 496 dots horizontally and coded by 24 bits per picture element—can be transmitted in about 7 seconds from the DVCP to a 64Kbits/s digital terminal.

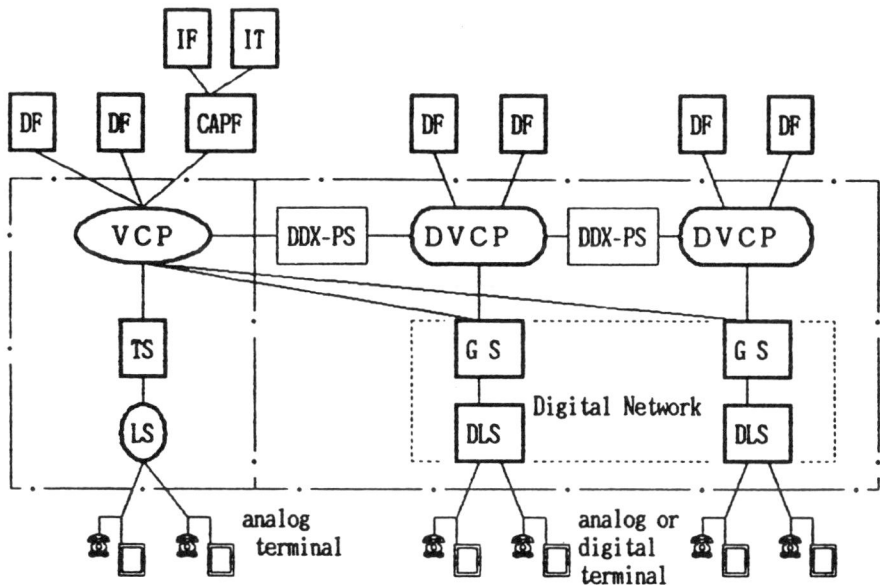

DVCP : Digital Videotex Communication Processing Unit
GS   : Communication Processing Gateway Switch
DLS   : Digital Local Switch

Configuration of Future CAPTAIN system

Figure 4.5

The digital terminals will be able efficiently to transmit mass data, such as information frame data created by digital terminals, to the information centres through the digital CAPTAIN network using a high speed upward path at a transmission rate of 64Kbits/s. Using this network capability, information centres will be able to provide various enhanced mailbox and bulletin-board services to digital terminals. Many types of messages will be possible: a character message which consists of alphanumerics, Kana and Kanji, a graphic message which consists of photographic, mosaic and geometric data, a hand-down message which consists of encoded lines, to name but a few.

Private digital CAPTAIN system

Besides the CAPTAIN system being used as a public videotex service, private CAPTAIN systems provide videotex services to users in local areas. Private CAPTAIN systems utilise a combination of existing communication facilities, including the PSTN, private branch exchange (PBX), and leased circuits, to connect user terminals to information centres. The information stored in the private CAPTAIN system is restricted to limited geographical areas or to special users, and its one information centre is generally small. Another feature of private CAPTAIN systems is that they adopt the same protocols as the public system. This means that the private centre can easily offer services nationwide by connection to the videotex communications network as an external information centre.

An advanced system of private CAPTAIN has been developed, and started operation in May 1987. The configuration of this system is shown in Figure 4.6. It is composed of a database management centre. Picture Response Equipment (PRE), user terminals that reproduce various received data, and input and editing terminals that create display frames to be stored, such as text, graphics, full colour pictures and audio information. The PRE is connected directly or through a digital switched network to user terminals and input and editing terminals. The principal services of this system are information retrieval service, multi-address delivery service, order-entry service, message board service, and gateway service (to other analogue videotex centres).

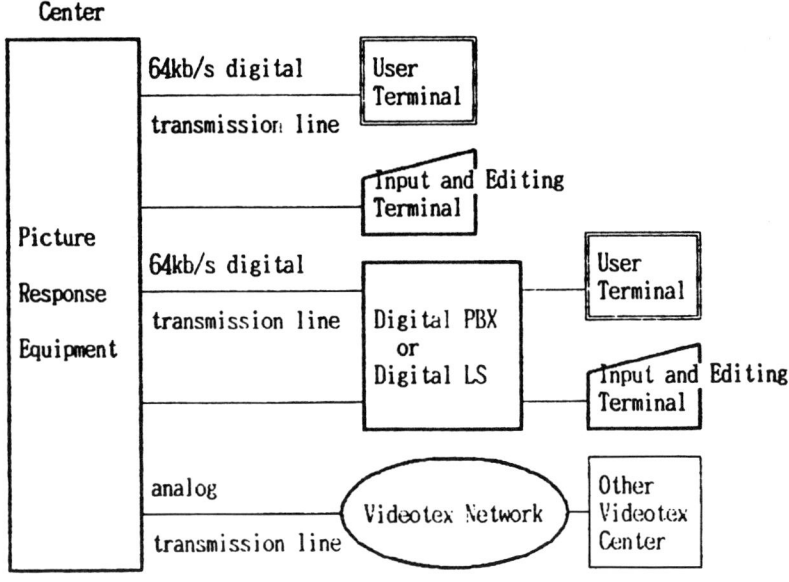

System configuration of Private digital CAPTAIN system

Figure 4.6

Conclusion

This paper has described the present situation and future trends of the CAPTAIN system. To make the CAPTAIN service more popular, NTT has been studying various enhancements to system functions beyond conventional retrieval services, operationality and information contents. In order to enhance system functions, it will be necessary to develop new services, such as transaction services (all sorts of reservations, mailbox, tele-software programs), information about the availability of different types of databases, VANs for many different industries, and connection to other networks (LAN, computerised communication network).

It is important to determine future user demands and to develop new technologies to offer services which satisfy these demands.

5

Videotex in France: the success of Minitel

Claude Finzi

The development of videotex services in France

After radio, telephone and television, a new medium of communication was launched in the early 1980s 'the Minitel'.

The first decision, taken by the French PTT in 1978, was to install an electronic telephone directory gradually to replace the paper directory, representing about 40,000 tons of paper each year, and to reduce the number of people employed by the enquiries service.

The main idea was to combine two technologies, data processing and telecommunications, and to use a dedicated inexpensive terminal linked to a database through the network Transpac. The advantages of such a solution were to provide information in all areas which would be automatically updated. The French government authorised the PTT to provide the consumers with either a free Minitel or a paper directory. However, if the Minitel was then only used for the electronic directory, it would have been a very expensive investment for the PTT. Therefore, the PTT took an ambitious risk, hoping for the development of new services based on the use of the Minitel which, as a consequence, would increase the traffic and hence the charges invoiced to the consumers. The major problem was how to solve the 'chicken and egg' question: consumers and companies have no reason to buy or to use a Minitel if services offered are not valuable and hosts will not invest until enough potential customers are available. In order to ascertain the reaction of the consumers to the use of Minitel and to assess which services would be successful, an experiment was launched in 1981: 2500 terminals were provided to a sample of volunteers in Velizy, a suburb of Paris both to test services and to find which services were of real interest to the user. It appeared that games and electronic mail offered by the local newspapers were the most popular, before the more practical information services such as TV or cinema programs, timetables and mail order. At the beginning of 1982 a new experiment, conducted by a newspaper *Les Dernières Nouvelles d'Alsace*, dedicated to mass-market users showed that 10 per cent of the traffic was for general information and services, 50 per cent for electronic mail and 40 per cent for humour and games. At the same time, professional time-sharing companies or hosts such as Questel or GCAM developed software to offer access to their information service by the Minitel.

The professionals (such as host services), and semi-professionals (services provided by in-house data processing services to closed groups of users) increased their usage progressively and regularly. Mass-market usage expanded rapidly, with information publishers such as *Le Parisien Libéré, Libération*, or information providers such as Funitel (games and entertainment). However, it was in 1985 that the mass-market exploded with the introduction of the 'Kiosque' concept. Before the Kiosque started, users of Minitel were obliged to know the Transpac address of the service they wished to use and obtain a password for database usage, in the same manner as if searching on-line. They received two bills: one for network usage, another for database usage. With the Kiosque concept, a user has to know only the mnemonic of the service needed. The bill is automatically sent by the PTT at a rate of 58.4 FF per hour, out of which two thirds is paid by the PTT to the host. This is much less complicated for the host; it is also less complicated for the user, who has the same user-friendly interface standardised access with any service used.

Some figures on Minitel usage

France's Teletel is divided into three areas.

Télétel 1 (3613) is aimed at providing professional information to professional users (such as Questel, GCAM and other hosts). The user receives one bill from the service, including network usage and database usage.

Télétel 2 (3614) is reserved for semi-professional usage. Network usage is paid by the user on his telephone bill at a rate of 22,2 FF per hour. The database usage is defined by the producer.

Télétel 3 (3615) is the Kiosque service, invoiced by the PTT at a price of 58,8 FF per hour; 36,6 is paid to the service and 22,2 is kept by the PTT for network usage and invoicing.

New Teletel numbers are planned in the future with different charge levels, from 37.5 FF per hour to around 500 FF per hour.

The 'Kiosque' evolution and trends

In an issue of *Information World Review*, George Anderla, former General Manager of DGXIII at the CEC laments "the walls of the capital have seen the flowering, over the past few months, of publicity placards displaying a great number of sexist or pornographic Minitel messages" (*Monitor, May 1987*). If it is agreed that more than 70 per cent of the usage on the Kiosque is concerned with sex, games and encounters and a new service is required, the investment needed to launch a new service today is enormous. Today's services must be more sophisticated and more inventive, and thus more expensive to set up, to develop, and to exploit.

At the beginning of the Teletel service, the pioneers were either Service Bureau companies with data-processing expertise, organisations such as newspapers with the information to put on-line, or companies for which information is a tool (timetable and booking for Transport companies, or Tele-ordering). These pioneers used Teletel 1 or Teletel 2. The situation stayed unchanged until the development of the Kiosque. When the Kiosque was installed by the PTT, and nobody knew what cost what, it was decided that the revenue should be shared equally by the PTT, the host and the information provider.

After mid-1985, a price war developed between hosts, and the percentage revenue share raised to 80 per cent for the producer of information, with 20 per cent for the hosts. Many information providers had decided to set up their own host: initially those like *Le Parisien Libéré* and later, after discovering that the host earned more money than themselves, like *Le Nouvel Observateur* or *Libération*.

A census of the traffic in May 1987 produced the following league table, with the number of hours of *usage per day*. All these services offer electronic mail, encounters, games and humour, telemarketing and information services (for example for sport and cinema).

SEGIN	: 9000	AZ Télématique	: 7500
SYTEM	: 9000	Le Parisien Libéré	: 7150
Le Nouvel Observateur	: 8000	C T L	: 6000
		FMC	: 4000

	1/3/85	1/6/85	1/1/86	1/6/86	1/1/87
Number of Minitel installed	614,000	742,000	1,217,000	1,724,000	2,131,000
Traffic on					
Teletel 1	435,000	232,000	431,000	480,000	545,000
Teletel 2	264,000	356,000	469,000	828,000	1,403,000
Teletel 3	736,000	1,033,000	2,201,000	3,467,000	4,698,000
TOTAL (hours for two months)	1,435,000	1,621,000	3,101,000	4,776,000	6,647,000
Numbers of mnemonics					
Teletel 1	327	350	515	602	725
Teletel 2	460	650	1,032	1,478	1,992
Teletel 3	57	160	352	752	1,435
TOTAL	844	1,160	1,899	2,832	4,152

Notes:

compared with   In 1986: 287,508,000 calls for 30,262,700 hours of connection, 1985: 103,974,000 calls for 11,679,600 hours of connection.

For services installed on Teletel 3 (Kiosque), the PTT paid back 822 million French Francs in 1986 against 278 in 1985 and 17 in 1984.

The number of calls on the electronic directory was 20 million, corresponding to 760,000 hours of connection in December 1986.

The evolution and trends of the Kiosque

Figure 5.1

Even for the industry leaders who have made a lot of money and have built a good reputation, launching new services is not easy. For reasons of convenience, of perhaps only because of the lack of know-how, they are searching for new partners with the required expertise or funds. SYTEM, CTL have just opened their capital to financial companies. Other mergers, are planned, bringing together bureau companies, TV companies, network operators and banks, to combine dataprocessing expertise, advertising support, and an appreciation of the needs of the users.

The concept of 'Telematic chain' is now a reality. Other services are becoming more and more sophisticated, and professional. The different levels of the Kiosque will allow service providers to offer a new range of services. I am able to conclude this discussion on kiosque services by predicting that successful future services will be provided by *managers* who know how to develop aggressive partnerships which will provide the services the users demand.

Professional usage of Minitel

With an installed base of three million terminals in 1987, an increase of revenue of 500,000 FF per year, and user-friendly access to the information requiring no training, it was obvious that the Minitel would tempt companies to expand their services and to give access to their internal information to a larger audience. A first result of Teletel has been to enlighten the on-line systems providers and prove that a service, easy to use and trouble-free, can attract a large number of users, and not stay in the 'ghetto' of the computer specialists.

Télétel 1 is dedicated to professional usage. The Minitel is a new terminal, for which companies have developed a user-friendly interface. A host like Télésystèmes Questel now has over a third of the searches made through a Minitel. A new service Questel-Entreprise has not been as successful as expected because of the need for a contract (to allow invoicing), and a password, and it has to be sold like the other services. It is, however, expected that the new levels of the Kiosque will allow hosts to offer databases through the Kiosque and increase their usage. Télétel 2 is dedicated to 'semi-professional' usage. It is interesting to point out the increase in the number of services (mnemonics) offered, which number more than 2,000 today, which coincides with the increase in the number of mnemonics offered through the Kiosque. It is impossible to describe all the services but three main categories stand out: enquiries services (including transactional services such as telebanking and teleordering); communications (directories and electronic mail); services which widen in-house telematic services. The two first categories are self-explanatory; the third is worth developing further with some examples.

Because of the high price of terminals, most of the original applications were limited to a small group of users. The Minitel widened the number of users without, however, modifying the application software. A company, with some synchronous terminals in its main office for example, provided all its vendors or stores with a Minitel. The result was an increase in both flexibility and commercial efficiency. All French banks are now offering services, to both companies and clients. After the initial experiment based on the on-line display of accounts and credit computations the services are now more sophisticated, covering cash management, evaluation of a portfolio, accounts transfer for individuals, all kind of credits, leasing, discounts, and loans. The list of applications continues to grow. A quick look at the 'miniguide of services' on-line with Teletel gives 22 themes such as education, economy, banking and finance, insurance, and health services. These are divided into sub-divisions and can generally provide more than 10 answers for each query.

Although an installed base of terminals is a necessary condition for the acceptance of a service outside a professional élite, the service must be easy to use, requiring no training, in addition to being trouble-free.

Teletel and new technologies

Information can be provided either as a print product, on-line, or on optical discs such as CD-ROM. The installation of more than 2.5 million Minitels, providing the possibility of searching more than 5,000 services, has reduced the impact of a number of alternative methods of information provision for two main reasons. Firstly, the Minitel terminal is cheap compared with the micro-computer; database usage is also generally inexpensive, if not free, for professionals. Secondly, the use of a Minitel is new, and to promote other methods of information will not be easy. These two factors explain why CD-ROM products have not had the same success in France as in other countries such as England, Germany or America.

A professional organisation *Le Cercle de la Librairie* published catalogues of books printed in France along with an ordering capability. Since 1984 a database of these catalogues has been set up with an electronic ordering capability; 1,000 subscribers are already customers and by the end of 1988 more than 1,000 publishers and 5,000 booksellers are expected to use the service. The main advantages of the service are the daily updating of the database and the decrease from five to two days of the sale cycle of a book. The cost of the equipment (Minitel, light pen and memory) is around £1,700. The price of an order is £0.004 with a minimum of £60 and a maximum of £1,000 per year. With the presence of such a service, the future of CD-ROM for booksellers in France is difficult to predict.

A similar system is installed for drugstores, and, of course, the 'yellow pages' of the telephone directory is already on-line at a cost of 22 FF per hour.

Taking the example of books further, an on-line system can be compared with a library, and a CD-ROM to a book. If someone wants information, in the same way as asking a librarian, he will search on-line. A CD-ROM, on the other hand, can be viewed as a book, providing text, images, drawings and charts. Depending on the frequency of updates, specific needs and prices, a CD-ROM and a Teletel service can co-exist.

Conclusions

To summarise, the main conclusions concerning Teletel services are that an installed base of terminals is a necessary condition of the acceptance of services outside computer specialists; an installed base of terminals is not sufficient to measure the success of a service which must be easy to use (no training, no contract) with a user-friendly interface; the number of terminals and services may have an impact on the development of new technologies such as CD-ROM.

6

Implementation of interactive videotex in the West German & British travel industries

Margaret Bruce & Falk von Bornstaedt

Introduction

The travel industry is essentially an information-based industry. It supplies a complete range and mixture of travel, accommodation and tour offerings as packaged or specialist holidays as well as offerings for business travellers. Consequently, travel offerings have to be packaged in ways that are attractive to customers and an effective information and communications infrastructure has to be in place to achieve the movement of people and goods from their points of departure to their arrival destinations across the globe. The communications system between principals and retail outlets, the agents, has undergone significant technical changes since the late 1970s, particularly with the development and installation of videotex systems to facilitate booking and confirmation procedures. Information technology has become such an integral part of the communications process that it is practically impossible to survive and attain a sustained competitive advantage as principal, carrier or agent without investing in automation.

The travel industry is a major service industry and a major user of interactive videotex in both West Germany and Britain. In this chapter, some of the major applications of videotex are outlined, together with a discussion of the factors influencing the adoption of videotex in the travel industry. Finally, speculations are made about the future role of interactive videotex in this industry.

Terms and background

Bildschirmtext, more commonly called 'Btx', is West Germany's public videotex system. It is operated by the German PTT, the Deutsche Bundespost. Bildschirmtext uses the CEPT-standard. It was launched as a national service in 1983 and now Btx has more than 100,000 users with a rate of more than 2 million accesses per month.[1]

Prestel is the national and public interactive videotex service launched in 1980 and operated by British Telecom, a private company since 1984. Prestel has 74,000 users, the majority of whom (59 per cent) are business subscribers.[2]

Interactive videotex and the West German travel industry

The nature of the travel industry makes it suitable for Bildschirmtext. Travel agencies are widely distributed and they sell a volatile product, one which cannot be stored. The travel agencies need up-to-date information about changing availability, conditions and prices of travel offerings. The need for up-to-date information is becoming more and more important in a deregulated and intensely competitive marketplace. Bildschirmtext is an appropriate technology to distribute travel offerings from the wholesalers (the principals and carriers) to retailers (the agencies). The information services on Btx are accessible from all parts of Western Germany on a local call basis and this gives travel agents access to many and an ever expanding number of reservation systems.

According to recent statistics of the Institut für Bildschirmtext und Telematik there are more than 500 information providers from the tourist sector publishing travel services via Bildschirmtext. These are: 116 tour operators; 167 tourist boards; 170 accommodation/restaurants; 44 transport/traffic; 30 others. An additional 40 on-line information and reservation services are available through the Bildschirmtext gateway system, half of these systems are provided by tour operators.

In Western Germany there are some 20,000 travel agencies, most of which are medium-sized businesses and, of these about two-thirds conduct their travel business as a side line. One in four travel agencies have on-line terminals, although the majority of companies with travel as their main business activity have on-line terminals. By the end of 1987, about 4,000 travel agencies were using Bildschirmtext. It can be expected that in the next few years those companies with travel as their main business activity will be equipped with Bildschirmtext and even a considerable part of the rest.

Information systems

The main information and reservation system in Germany is START. More than 2,500 travel agencies are using about 5,000 START terminals today. START was introduced in 1979 by the national airline, Lufthansa, and national railway operator, Bundesbahn, and the largest tour operator Touristk Union International (TUO). START gives direct access to the reservation systems of these carriers and tour operators. In addition to its information and booking facilities, START offers comprehensive accounting facilities for travel agencies. START works on dedicated lines for data transmission and give the comfort of high transmission speed and reliable functions on an ergonomic screen with 80 characters per line. START printers recognise more than 30 different types of documents like airline tickets in various formats and thicknesses which they can print immediately. Since START was relatively expensive, it was mainly used by the larger travel agencies until about 1983.

In 1983 Bildschirmtext was introduced as an official service of the German PTT on a national basis. From the beginning, the travel industry was heavily involved in Bildschirmtext, but not quite to the same degree as in the UK, because of START and its competing on-line systems. Ironically, START even contributed to the success of Bildschirmtext by offering access to its system via the Bildschirmtext gateway. In 1985 START succeeded in developing a configuration with a personal computer, which can be used like a normal Bildschirmtext terminal with 40 characters per line or like a START terminal with 80 characters. This low cost solution for using START was a big success, nearly 600 travel agencies are using it, especially for booking TUI-tours.

In the beginning, START was a closed shop: the stockholders had no interest in opening their system to other tour operators. The only threat they had was the development of competing systems. However, the costs of setting up an alternative to START were so high that this did not constitute a real threat. This situation changed with the emergence of Bildschirmtext, because reservation systems could then be established at much lower costs: consequently, START lowered its barriers to entry. Today, more than 30 service suppliers in the travel market have concluded contracts with START. Nevertheless, the videotex solutions are not quite as good as the classical on-line terminals, but they are sufficient for many applications.

Even the users of START were not allowed to use all parts of the system. Eventually START operators were forced by a German court to open up the flight information section to a travel agency; the court stated that START had violated German cartel legislation.

Bildschirmtext helped the travel agencies to become more independent from the larger tour operators. It is an open system enabling travel agents access to a number of on-line reservation systems at a relatively low cost.

Looking ahead

Bildschirmtext has not yet entered the private market, only about one in four users are private subscribers. In the 1990s it is expected that Btx will have penetrated the markets for private households on a large scale. Then it may be profitable for companies to offer a database of package tours designed to be used by both travel agents and householders. There is already one database in Germany giving this kind of information.

Videotex also enables wholesalers to offer direct bookings to customers without intermediate travel agencies. In the years to come, there may well be more and more direct booking facilities. On the other hand, travellers will still need assistance from travel agencies to guide them, for example, through the array of different tariffs.

It is possible to retrieve travel information from Btx from home, but, as yet, many private users do not have the skills to use such databases effectively. If procedures for the use of computers becomes easier, then household users should be able to reveal their travel preferences to a computer and may even be able to obtain better information than from their travel agencies. In this scenario, the standardised packaged mass travel business might be lost by travel agencies. However, for 'off the beaten track' and for individualised travel plans there will still be enough business opportunities for good travel agents. Nonetheless, those agencies who continue to see their main function as the distribution of travel prospectuses and the handling of bookings will find their existence increasingly threatened.

Interactive videotex and the British travel industry

The travel industry has been a success story for interactive videotex. Within five years of its introduction to the industry late in 1979 about 90 per cent of all UK travel agencies used videotex systems for checking late availability, up-to-date prices and general travel information, such as currency exchange rates. Over 85 per cent of all package holiday bookings are now made through videotex. As well as the travel services available through British Telecom's Prestel service, there are private videotex systems designed and operated by principals and carriers. The most famous is perhaps that of Thomson Holidays' TOPS system and the videotex networks provided by Istel and Fastrak. Videotex technology has become an integral part of the communications infrastructure of the travel industry, so it is not possible for any particular group of the industry to work effectively without interactive videotex.

Videotex has been used mainly to improve communications between the tour operations and travel agencies. Why has interactive videotex been taken up so rapidly by all sectors of the UK industry? Up to the 1980s, the main communications channels between the carriers, travel principals and the geographically dispersed agents about travel offerings had been via telephone, paper and telex. Booking and confirmation procedures were time consuming, with loss to potential custom, and travel companies were unable to realise the full value of their products. Interactive videotex offered significant advantages for improving the distribution of travel products, in particular, information about availability of airline seats, hotel accommodation, and the opportunity of making electronic confirmation of reservations.[3]

The travel principals and agencies initially adopting videotex were able to gain a competitive advantage by reducing the cost of communications between principals and agents, for example, with the reduction in the production and postage of written information and by improving communications, especially in the dissemination of up-to-date information about their travel products, such as prices and availability. The service to the consumer improved too, resulting in increased custom for those companies with electronic systems.

Videotex applications

The companies taking up interactive videotex exploited its potential in different ways. One effective strategy was that of Thomson Holidays. Initially the company designed and developed its own private videotex system, TOPS. This restricted access to Thomson Holidays' travel information to those agents trained to use TOPS, which ensured that its travel information was easily selected by the agent selling its travel products. It also meant that the principal's information was secure; its competitors could not find commercially sensitive information about, for example, the company's load factors. TOPS also led to cost-savings in the communications between Thomson Holidays and the agencies which distributed its products. Then, in December 1986, Thomson Holidays made all its holiday bookings available via interactive videotex only. Indeed, during 1986, the company doubled its business, increased its market share from 20 per cent to 30 per cent and kept its average holiday prices at the 1984 level. The company attributes this success to the effectiveness of its TOPS videotex system.

As regards the retailers or travel agencies, the adoption of videotex is perceived as being advantageous in being cost-effective (mainly because a greater volume of business has been handled by the agencies without incurring substantially greater costs) and by making the process of booking, confirmation and finding the latest availability of prices of travel offerings so much easier.

These benefits have resulted in improved customer service with the implication that the agencies with videotex gained a larger market share. However, the interactive videotex wave of technological change in the travel industry is nearly over. Videotex has become one of the industry's main standards and has led to the automation of the travel industry. Now companies are looking for new technological changes to exploit.

Speculations

The main technology-based developments most likely to substantially affect the industry are a range of value added networks and services (VANS) but operating in a non-videotex environment. Key players in the travel industry are collaborating to supply such VANS services. Although competing VANS systems are expected to impact the industry in the next couple of years, the services they provide will be broadly similar. Electronic reservations and confirmation systems will be on offer for 'composite' bookings (such as package holidays with flights and accommodation) being delivered as an inclusive package deal. However, non-composite offerings will also be available. The latter allow for separate reservations to be made for flights, accommodation and so on. In so doing, new permutations are possible for agencies to build up a customised offering to match exactly individual travel needs from the available travel offerings.

The trend to 'self-service' selection and bookings, happening in West Germany, is occurring in Britain too. One indication of this is the increased use of public-access ticketing for flight and train bookings. Putting aside the political issues surrounding direct bookings—in particular who will win and lose with such a development, the principals (wholesalers) or agencies (retailers)—the travel consumer would have to be confident that the databases were comprehensive, accurate and reliable and that liability is guaranteed if something went wrong.

Human effects of automation

It is possible to discern certain trends with the use of videotex in the British travel industry. These include job loss for those employed in the reservation and information offices of tour operators and carriers, and a move towards centralisation and rationalisation of the industry as the communication networks and distribution outlets are structured around information technology.[4] New jobs are also being created and new skills are required, such as keyboard and information retrieval skills, learning how to sell travel products whilst using the technology and how to make bookings, to name but a few. Training and education for these new skills is addressed by private companies, and managers of travel agencies and some colleges have an information technology component. However, the provision and quality of training and education varies.

As with West Germany, travel agencies are concerned about their survival in the longer term since videotex offers opportunities for travel principals and carriers to sell their products directly to domestic videotex users and to bypass travel agents. This form of distribution of travel products may need changes to laws of product liability and improvements to the information search-and-retrieval procedures. Agencies may have to change from high street booking shops to travel consultancies catering for individual travel requirements and tastes.

Conclusions

This comparison of videotex applications and developments in the West German and British travel industries reveals some similarities and differences which influence the rate and pattern of adoption of interactive videotex. The technology offers relative advantages for communication networks and distribution outlets for travel companies in both countries.

There is the issue of 'who controls' (the principals and carrier or agencies) the design and development of the technology and what effects 'who controls' may have. This issue is relevant to any consideration of the opportunities videotex provides for distributing travel products to the domestic videotex user.

The different structures of the travel industries in West Germany and Britain, the different public videotex systems provided by the PTTs or system operators and the pricing policy of suppliers of videotex and alternative electronic systems available to the travel industry all affect the adoption of videotex by the travel industry in the two countries.

For the longer term, in both countries, the type of and quality of training and education for staff in the industry needs to be addressed, as well as technical considerations regarding design of information retrieval procedures, the use of microcomputers with access to videotex systems for on-line reservation, accounting and client filing systems and the legal situation regarding product liability.

The rapid and widespread adoption of interactive videotex in the travel industry has paved the way for the newer VANS systems to be implemented. This will again fundamentally alter the industry.

7

Videotex publishing: the US experience

Christopher Burns

It is over ten years now since the British Post Office first showed Prestel to outsiders, and a great deal of time, enthusiasm and money has been spent to test the hypothesis that this technology was, as Rex Winsbury put it, "push button publishing".

Alex Reid, then head of the BPO Prestel Division, expressed the hopes of everyone when he wrote in, 1979, "Our aim is to produce a cheap and universal means of electronic publishing, available to all; a new medium of communication, comparable in scale to radio, television or the press . . . As with any new product," he cautioned, "there is a risk that the customers will not take to it. We have, in any case, a gut conviction that this kind of service will come."

Today we are faced with a remarkable dichotomy of experience. According to George Nahon, managing director of Intelmatique, the French PTT earned nearly $70 million in revenues during 1985 from the increased telephone usage, cost savings and terminal leasing fees associated with their videotex system. On the other hand, a number of US companies have invested at least that much without finding any significant revenue at all.

In 1986 Times Mirror announced the termination of its consumer videotex effort, Gateway, with the terse observation "The reaction among consumers, while gratifying, was not sufficient to warrant full scale development of Gateway as an on-going business". The same year the final vote on consumer videotex in the United States was cast by Knight Ridder which closed down its Viewtron experiment after having invested as much as $50 million over five years.

How can such views be explained? Perhaps the French success is merely the fad popularity of 'Chat' on a large scale. Every videotex experiment has experienced an early rush of enthusiasm for the anonymous exchange of messages—a sort of electronic graffiti that begins with innocent (but usually illegal) offers to swap software and gropes ever more boldly toward the prurient limits of the hacker's imagination. Perhaps, by making the terminals practically free, the French system has pushed videotex to critical mass. Like the telephone system in its early days, usefulness grows exponentially with the number of users. Or could it be argued that, by making directories a basic part of the service, the French system placed its Minitel within that limited set of information tools we have to use every day? As the terminal became familiar, the obstacles to using it for more novel purposes fell away.

Whatever the secret ingredient is for success (or the appearance of such) in the French system, it is not pushbutton publishing as it was once imagined. Whatever strengths videotex may have as a transaction system, a retail outlet or a message terminal, even the Minitel has not been successful as a means for distributing news or permitting research. That raises some interesting questions about future opportunities for publishers in this medium, and indeed about the nature of publishing itself.

Did we ever really try publishing?

In *Goodbye Gutenberg*, Anthony Smith came as close as anyone to presenting the thesis that videotex would displace the newspaper. And Heaven knows we tried. No one invested more in this attack on the newspaper than the publishers themselves who, singly and in groups, tried every known combination of headlines, helpful hints and hard sell. In his heart, each publisher feared that videotex might break out in his market the way television did thirty years before, sapping his advertising revenue and challenging his dominance in the news business. In retrospect, there were serious flaws in most of these experiments. On Compuserve, many of the major newspapers routinely embargoed their news for 24 hours before putting it up on the system, thus ensuring that it would not be competitive with the print product.

In other cases, editors carefully shaped the electronic service for a mass audience that was not there. Not until late in the game did we acknowledge that the only persons who could access electronic publications were computer hobbyists, analysts and planners trying to track the medium and editors of other electronic publications. Even in 1984 one of the heaviest users of Viewtron turned out to be the videotex development team of a competing company.

In 1980 the American Newspaper Publishers Association threw its considerable weight across the path of AT & T, and brought the Albany and Austin electronic directory programme to a halt. The near-term prospect of losing classified revenue, coupled with a long-term concern for open competition in this medium convinced the publishers, and Congress, that telephone companies should not be permitted to become electronic publishers. After The Breakup, of course, the point was lost. AT & T tried earnestly to find a way to electronic publishing but it now seems to have abandoned the idea.

While each of these issues slowed the development of successful systems, none of them were, in my opinion, mortal. On the contrary, looking back at the last few years of videotex, one is struck by how thoroughly the publishing experiment was pursued.

Every which way but free

The tailored newspaper?

From the outset, The Source, Compuserve, Viewtron, Gateway and others stored the wire service in a searchable structure so that individuals could scan for stories that served their interests. Technically it worked. But, although the AP file was the most popular service on the Compuserve experiment, the access schemes were discouragingly cumbersome and the search mechanism was little used. In business videotex systems, though, the ability to scan *abstracts* of stories has been a consistently popular feature.

The videotex access scheme is a major issue. Because of the way most experiments grew, a structured menu was the normal tool for selecting the information, and that very quickly became tedious. Several systems were able to move to keyword searching, but neither Gateway nor Viewtron, the most fully developed of the services, was able to move to software that would recognise the subscriber and anticipate his needs. In this respect, the experiments fell short of building what was possible, choosing instead the inexpensive but very inefficient menu/page structure.

Moreover, the original vision of a tailored system was uneconomical from the start. Prestel quickly found that when one expanded the scope of coverage greatly and spread the cost of each item over fewer customers, the price to each reader was simply too high. There were no economies of scale and they quickly had to cut back to areas of coverage where there was broad group interest.

Timeliness?

Every experimenter sought at one time or another to provide instant news, and in a number of cases videotex systems were able to put important wire service stories into the hands of readers within moments. While this was pleasing to the experimenters, it went largely unnoticed by the subscriber who did not realise how fast the story had been put up, did not care, or signed on only at night anyway because the rates were lower or because that was when he was home.

Local detail?

At one point Viewtron assigned six reporters to gather the local news about a community of 25,000 persons—in addition to all of the news normally available about that community from the *Miami Herald* staff. For months this effort went on, reporting every community event, sports score and issue that seemed useful. Subscribers paid no attention.

Special interest information

Throughout the five years of experiments in the US, services which focused on the interests of a narrow users group seemed to do well, particularly if the users group was computer hobbyists. Investors, doctors, software designers, games aficionados and telecommunications managers found that electronic publishing was faster by weeks or months than the trade journal that served their narrow group. But their needs were occasional, not daily, and usage was low. By its nature the information was marginal and subscribers never developed the system loyalty (or 'dependence') that is necessary for success.

Cost

In retrospect it seems odd that it could ever have been imagined that videotex would cost less than newspapers. In 1980 Anthony Smith reported the then prevailing view that videotex would cost the average household less than $100 per year. A printout would be pennies. The cost of becoming a publisher would be a few thousand dollars for editing and frame creation equipment plus a few pennies per page per year. The electronic delivery system, of course, would be quite cheap since it did not consume newsprint or other costly materials. But things did not turn out that way.

In the last few weeks before stopping the Viewtron program, Knight Ridder began an introductory campaign that amounted to free videotex for a few weeks. Subscriptions increased and usage went up dramatically. Macro-economists would respond that newspapers had set the acceptable price of information at 25 cents per day (Television sets it even lower) and that a competing videotex service would be viable only when the price dropped below that level.

Do-it-yourself publishing

Much was made of the impact videotex could have on society. Ted Nelson, author of *Literary Machines*, observed that before the printing press copyists often added comments to the text as they copied it and even modified the text to suit their own views. For any major work there were many versions. With Gutenberg, we moved to a concept of standard editions to which everyone referred. Videotex, he argued, would liberate us from such restrictions and we could once again have many reports of an event. Whatever biases or omissions the media establishment may be guilty of could now be remedied. As the printing press disintermediated the church, so videotex would disintermediate the press.

Except it did not happen. Initially the bulletin boards attracted individual accounts, stories, poems, propositions and other moments of literary flight, but it soon became apparent that no one was reading this stuff. By contrast, for all of its conforming influences, the press delivered information in an even-handed and sometimes attractive way.

Videotex continues to thrive as a message system, and at some point on the continuum, messages become reports and reports become publishing. But in retrospect we now see that the vision of all these individual 'publishers', tapping furiously at their keyboards, contained an important flaw: readers value the role that publishers play in editing text, checking its accuracy, presenting it effectively and putting an idea in the context of other thoughts on the subject.

Did videotex get a fair trial?

Videotex did get a fair trial. With some reservations, the US videotex experiments tried every combination of publishing services imaginable. As a publishing medium it does not work for reasons that seem not to be technical, although the access scheme was crude. A major problem is certainly our tendency to overestimate the value of information to consumers, and to underestimate both the power of the print medium and the economic resilience of the news/advertising coalition.

In the last week of Viewdata's operation John Woolley, Viewtron's designer and editor, was asked how he would do it differently if he had to. He said he would build a videotex system that called the subscriber. Perhaps we should move closer to an intelligence system, he suggested. A weather alert, a tip sheet, a gossip network might be attractive. We may be crossing a boundary, he suggests, from the broadcast of useful information that has characterised the last 500 years of publishing (where videotex does not work) to something more like an information network that calls you, like a loyal friend.

Even the name betrayed our view that it was a faster, cheaper way of *distributing content*. We imagined a vast but browsable library of knowledge, a network that delivered instant news with comprehensive detail to back it up. But in reality the speed, details and selectivity which was imagined for electronic publishing happened only at great cost, and videotex almost immediately became a message system. At least that has been the lesson of US experiments so far.

Are they any changes abroad that will cause us to return to videotex as a publishing medium? Yes. Firstly, over the next decade we will find that most of the information bound for publication becomes accessible by computer at an increasingly early stage. The files we now rekey for a database may in the future be directly accessible. The cost of building and maintaining a current knowledge domain will be lower. Secondly, the current generation of radically new knowledge management schemes, while themselves impractical, promise intelligent and pragmatic offspring. The sign-on and the menu structure access will quickly fade from view, to be remembered with a smile from the high ground of expert systems and heuristic software. Thirdly, and most important, we can look forward to a better understanding of how consumers acquire, use and value information. Thus armed, we may be better able to provide an appropriate utility.

I think it is sad, but not surprising, that out of nearly $100 million spent in the United States on videotext experiments in the past six years, so little was spent studying the consumer's information acquisition habits that there is no trace at all of the work. Because we thought of publishing, we looked at publishing models (magazines, newspapers, newsletters, direct mail catalogues) whereas I think Woolley is saying that we should have given equal time to gossip systems, colleague networks and information processing teams.

Videotex was not a publishing system as we imagined it. Perhaps another decade of software development will bring us to the point of trying the experiment again, but for the moment we shall watch the evolution of the French system with enormous interest.

8

An introduction to electronic shopping: the coming revolution in retail marketing

Thomas R Rauh

In the past two years, a number of large companies have formed joint ventures to offer videotex services to households with computers. These ventures have been led by computer manufacturers, financial institutions, media companies, telecommunications and computer service firms. Although slow to get involved, retailers have a number of important reasons for pursuing the development of electronic delivery systems. Profitability in most retail segments has been on the decline since the early 1970s, as the trend shifted sharply towards self-service and no-frills shopping in general merchandise. The most rapidly growing segments since 1980 have been warehouse stores and off-price apparel chains. In addition, non-store retailing has proved more profitable than store-based retailing and has been growing faster. The mail-order industry has consistently outperformed other retail segments in sales growth and return on sales over the last five years.

These factors, combined with the conviction that electronic shopping offers higher returns, incremental sales, and improved productivity, have caused more and more retailers to take a serious look at interactive electronic communication with customers. They want to know more about what it is, what forms it will take, how they can gain or be hurt by it, and how and when it will become a major factor in the retail marketplace.

The forms of electronic shopping

Today, electronic communications for shopping revolves around videotex and videodisc. Videotex is a bundle of information available to subscribers through existing telephone or cable television hookups. Included is text and graphic information such as news, weather, sports, airline flight times, as well as interactive services such as shopping and banking. Systems are menu-driven, listing all of the information categories available. If the consumer selects shopping, a menu lists different types of shopping from which the consumer identifies the type of merchandise or retail store of interest. The consumer is then presented with an array of brands, models, prices and stores, and can go deeper into the system for all of the information needed. When the decision is made to buy, the consumer enters the charge card number and orders the item.

Videotex has a number of advantages over current mail order procedures. One advantage is that videotex can be updated quickly, almost instantly. Cruise ship lines in Miami, for example, offer 50 per cent to 75 per cent off available staterooms on the day before sailing, rather than have empty rooms. Theatres reduce ticket prices the night of a performance to fill the house. For retailers, this updating capability means announcing price changes instantly, depending on performance trends, timing, and other factors.

Videotex can also be used for comparative shopping. Consumer Reports, an early participant in at-home systems, sees the media as a way of providing a much wider audience for its ratings and product evaluations. As consumers use the videotex to cross-reference brands, models, and prices, they can also compare ratings before making a decision.

Videodisc is a means of storing digital information on a disc. Text-type information, photographs and videotapes, including sound, can all be stored on the same disc. Videodiscs will be used primarily in in-store and public-access terminals to assist customers in gaining product information. As much as 30 minutes of full-motion video and audio can be stored digitally on discs. To view the information, the customer activates the terminal, which calls up the appropriate products on the disc. Videodiscs have already been used successfully in selling high-tech products, particularly when it is difficult to train salespeople adequately. The discs inform consumers about microcomputer hardware and software, providing information that sales people often do not know, or forget. Discs can compare products and suggest alternatives and related purchases.

Videodiscs are produced at a central location and cannot be erased. Although production costs are high, they compare favourably with the cost of network quality television commercials. When used to display still-frame images, videodisc production may also be competitive with conventional print catalogue printing expenses.

Major markets for electronic shopping

Two major markets are seen for electronic shopping: the at-home market and the public-access market.

The at-home market requires a special terminal device that hooks a consumer's personal computer or videotex terminal to a computer center via the telephone system. The at-home market will probably develop slowly over the next ten years. The price for the special home terminal—$600 to $900—is more than many people are willing to pay right now for shopping, banking, and information services at home. In addition, at present these services can offer only text and graphics that are not very exciting to most consumers.

Once the presentations are improved and the cost of the terminal is reduced, and both can be foreseen, the acceptance of at-home shopping is likely to be very strong. Research shows that 45 per cent of consumers in the United States would subscribe to an at-home shopping service, and that 25 per cent would use it regularly. That is a significant set of numbers considering the capability of providing every home in the country with this service. In recent field tests, some general merchandisers have achieved broad market acceptance for at-home shopping, with as high as four transactions per household per year. The average transaction was $50 to $60. Market research indicates strongest consumer interest in at-home supermarket and drugstore shopping. One major advantage of at-home shopping systems is their ability to organise and present a large amount of data for shopping. This would reduce the time required for supermarket shopping, particularly for routine items and staples. At-home service also simplifies the process of comparing store and product information, saving the consumer time and money.

The public-access market involves placing terminals in airports, hotel lobbies, offices, factories, and shopping malls. People will have access to these terminals as they do to Automated Teller Machines (ATMs). In shopping they will be able to obtain information about stores and merchandise and, if they choose, to place orders through the terminal. These terminals may be capable of delivering television-type presentations through videodiscs, as well as text and graphics. Shopping terminals in public places can generate incremental sales for the retailer without brick and mortar costs. Public-access systems can be used to sell in marginal locations, in the workplace, and other areas where the people are but the store is not. They can extend store hours without increasing personnel costs.

These terminals will also be placed in shopping malls to help consumers find the merchandise they want more quickly. People can be aided by a number of menus on the terminal screen. One menu can present a listing of stores in the mall by alphabetical order, type, and location. Other menus will focus on merchandise by type, brand, price, and so on. Cross-references can also be provided. These terminals can be a means of advertising to increase traffic and sales per visit. A customer seeking information on an item may be shown information on related items, and may decide to add to the original purchase.

Terminals in the store itself can increase sales and customer service while reducing selling costs. In-store terminals can be used in high service departments to support less experienced or fewer salespeople. For example, Ford Motor Company has purchased 10,000 terminals to place in its dealerships around the country so that customers can see the 'infomercials' on the product lines they are interested in. These presentations vary in the kind of information provided. For example, the buyer interested in styling can be shown that, and a customer concerned with technical specifications can learn about engine details. Presentations in graduated degrees of detail can be made for household appliances, home furnishings, toys, sporting goods, and electronic and photographic equipment. One luggage manufacturer claims most salespeople do not know how to demonstrate its better-quality lines. The in-store terminal will do that more completely and effectively.

In-store terminals can also help make the sale when the item is not in stock. The customer can see the item on the terminal, make the selection, and place the order, knowing when it will be shipped. Retailers can provide a wider range of selections at lower inventory costs. For example, a retailer with in-store terminals can 'carry' 15,000 types of towels and sheets while having only 1,000 styles and colours in the store.

Roles for retailers

Retailers can play a number of different roles in these electronic information networks. The least complex is the information provider. That is, the retailer offers products and services to the consumer via someone else's system. The retailer pays a fee to the system operator to carry information about his store and products over the system. The system operator oversees the operation of the computer and the communications network. As an information provider, the retailer would typically incur start-up costs of $10,000 for videotex and at-home services, with minimum annual cost of $15,000. For a videodisc public-access system, the start-up cost is $50,000, plus $10,000 a year.

A more complex role is that of the internal system operator. The retailer places terminals in the store and in other public places. He owns the computers and develops the information presentations, probably using outside video production houses. While this arrangement involves more work for the retailer, it also offers the greater control and flexibility of a proprietary system. The potential for cooperative advertising with manufacturers also makes this role attractive. As an internal system operator, the retailer would have a start-up cost of $100,000 for a videotex gateway service, plus $50,000 a year. For text and graphics terminals the start-up cost is $5,000 to $7,000 per terminal plus $100,000 a year. For an integrated videodisc system the start-up cost is $15,000 per terminal, plus $100,000 a year.

The third and most complex role is that of the external system operator. Choosing this role, the retailer would actually go into the information systems business, delivering not only his own messages but also a bundle of services to market-driven locations. An external system operator would incur start-up costs of $10 million to $20 million for at-home services and $6 million to $10 million annual costs. Those costs could be cut in half for videotex public-access services.

Getting ready

There are a number of obstacles that must be overcome before electronic shopping becomes a major factor in retailing. The obstacles will be overcome; the only question is how long it will take. Electronic networks, particularly at-home systems, are very expensive to develop. The cost for at-home subscriptions is also high. Right now there are not many items on the systems, so people cannot justify the subscription cost. Advertisers and information providers will not go on the system unless there are more subscribers.

Electronic shopping, though, is building momentum. The number of pilot programs and actual functioning systems, coupled with the quality names involved, point to a growing acceptance of electronic shopping in the next few years. Knowing how quickly the future becomes the present, many retailers are preparing themselves now for the coming revolution in retailing—electronic shopping.

9
Electronic shopping outlook: 1987–1992

Thomas R Rauh

By 1992, between $2 billion and $3 billion in sales will be generated on electronic shopping terminals annually. Those sales will be made on 30,000 transactional terminals throughout the United States. In addition, there will be over 200,000 information systems in public access and store locations. The home will be another location for electronic shopping, serviced by cable and interactive television services.

While these numbers may seem startling, people close to the electronic shopping industry consider them to be conservative. A host of retailers, manufacturers, and communications companies have been transferring their ideas, theories, and dreams of electronic shopping from the drawing board into reality.

The electronic shopping marketplace

Three broad segments have emerged in the electronic shopping marketplace: the home market, the public access market, and shopping terminal systems market.

The at-home market will reach shoppers in their homes. This market requires that a consumer's personal computer or special purpose terminals be hooked to a central computer that feeds shopping information into the home. This process is known as videotex. The link between source and receiver can be made over conventional phone lines. Through videotex, consumers will receive information in the form of text, graphics, and photo quality images. Other major avenues into the home are conventional cable television, broadcast television, and interactive video. Interactive video weds computers to television so that the viewers can call-up on demand the images and information they wish to see.

The public access systems are primarily informational services, delivered through terminals in a wide range of locations. These systems provide automated directories and other information services in underground walkways and shopping malls, building lobbies, and of course, in retail stores. These systems will carry a comprehensive overview focused on small databases with emphasis on businesses in close proximity to the terminal—the tenants in a building or mall, the products and services in a store. Public access systems should be dynamic, meaning they can be changed frequently and easily. Within a small geographic area, then, we will see a number of units providing different information to the consumer.

The shopping terminal market seeks to merchandise specific products and services and asks consumers to make a purchase decision on the spot. Location will be a feature of the shopping terminal market, too. The primary focus will continue to be the retail store, but there will also be free-standing kiosks in public access locations such as the underground walkway. The primary vehicle for these electronic catalogue systems will be interactive videodisc technology, emphasising photographs and voice-over stills. One of the most exciting possibilities for the shopping terminal is the ability to use videodisc and related technologies to upgrade the functionality of existing ATM systems. With perhaps two terminals in the same ATM enclosure, consumers will be able to view videobased presentations, make financial transactions, shop, purchase tickets, make reservations, and so forth. Shopping terminals may also find a welcome in workplace locations such as employee cafeterias. Particularly in areas where employees cannot get to a retail store during the day, the workplace terminal may become a popular means of shopping convenience.

Initially the most successful applications of transactional systems will be in selling branded merchandise off-price. That is a concept all consumers understand. When they know the brand, they know what they are getting. If they see that the price is lower than normal, they will not hesitate to seize the bargain.

These three marketplaces—at-home, public access, and shopping terminal—offer a dizzying array of locations that could be developed for electronic retailing. There are an estimated 500,000 suitable locations for public access terminals in the United States, and many millions of households potentially reachable through these technologies.

Strategies

How are companies approaching these markets? The focal point thus far has been in existing stores—to shore-up declining service levels. As retailers have cut back on their selling forces over the last 15 years, shoppers have found it increasingly difficult to get help from someone who knows the merchandise. The new technology is being used to put quality selling back on the floor.

One example is a system developed by Elizabeth Arden, which is running on an experimental basis in department stores around the country. The system allows a woman to sit in front of the video camera, and have a picture taken of her face. The picture is then transferred onto a computer screen. She moves in front of a screen and with a cosmetologist tries a variety of makeups and products on the image. She actually sees her face made up with each new combination, without ever touching any of it to her face. At the end of the session, when the customer has decided which look she likes best, a print-out is generated by the computer, listing all the products she has chosen. The customer takes the printout to the sales clerk at the counter who picks the products and rings up the sale.

Another interesting example is Eye Works, a division of Cole International, selling fashion eyeglass frames in many locations including Sears stores. Eye Works' problem is somewhat unique in that they have about 4,000 fashion eyeglass frames, in a number that overwhelms the average consumer. With the aid of expert systems software and interesting graphics, Eye Works developed a system whereby customers can pick out certain facial features which correspond to their own look. Based on their answers to short list of questions about themselves, the computer suggests four or five eyeglass frames for the customer to try. When this system is tested against the recommendations of experienced sales people, 60 per cent to 80 per cent of the computer recommendations agree with those of the sales experts. The results of the Eye Works system have been increased sales and improved customer satisfaction.

Besides employing the technology to aid salesperson productivity, retailers are also using it in the transactional market to sell directly from central stocks. Essentially the transactional terminals permit a mail order-type operation to be set up using the terminals as the vehicle through which merchandise is presented to the customer. There are a number of concepts people are moving towards in implementing these systems.

For example, Florsheim Shoe has a system that deals with the issue of depth in current lines. Their video system emphasises styles, sizes and colour that are not otherwise stocked in a conventional Florsheim store, but are available from the factory. Their success with this system has warranted an expansion of the programme. Another strategy focuses on selling complementary merchandise, such as customer electronics in a supermarket or convenience store. The converse of that—selling groceries in conventional retail stores—may begin to show up in the next few years. There is the big store/little store strategy by which retailers looking to expand into new markets reduce the cost of bricks and mortar through the use of this technology. New stores can be smaller and operating budgets lower, without sacrificing the merchandise offered to the consumer. Cross-divisional merchandising is still another strategy. There are many companies in this country with multiple divisions that have very different identities. The new technology would allow the department store division to offer merchandise from its speciality store group, and vice versa.

Finally, the current strategy for existing stores is to present merchandise to customers more effectively. Many products sold in retail stores today are not presented effectively, particularly more complex products, such as those requiring demonstration, those with multiple features, options and accessories. The technology can complement the knowledgeable salesperson and offset the lack of trained personnel.

The focus of the new technology thus far has been on existing stores, and this will continue to be a major location for shopping terminals. Perhaps the most important application of the new technology, though, will be in developing new types of stores and new locations not traditionally known for selling. The dramatic lowering of operating expenses must be seen as a major consideration for electronic shopping.

This array of strategies is likely to result in the development of new formats such as the electronic supermarket, the electronic speciality store, the high-tech, high touch combo store, and free standing kiosks located in walkways, cafeterias, and lobbies. But services are expected to develop in transitional phases. At first retailers may have a number of terminals at, let's say, the cosmetic counter, with each terminal sponsored by a different cosmetic manufacturer. As the number of terminals grows, and each has a different look and technical requirements cluttering up the selling floor, the retailer is going to have to take the initiative, bringing together the various product messages into one system offered throughout the store. It is not clear when this will happen, but it will.

Another direction that is likely for the electronic movement is the multi-store and the multi-service kiosk located in free standing stores and in public access locations. They will create a whole new business for the 'electronic mall operator', providing an umbrella under which retailers and service providers come together through this electronic kiosk or chain of stores.

The technologies

Much has happened in the development of electronic shopping, yet much of what is still to happen will depend on the progress and direction of the technology itself. There is still a long way to go. For example, the state-of-the-art in shopping terminal systems today would best be described as ATM-style small screen systems. Most of the systems are menu-driven with a sequential access method that is slow and difficult to get through. Typically these are single, free standing units. The videodisc production process for these units requires a long lead time, and is complex and costly. Because of the low volume, average unit cost for the hardware is high. A second generation is emerging. The newer systems utilise lower cost technologies such as LCD displays and digital audio. These systems, pioneered by Intermark Corp. in New York, are being applied to a wide range of product categories including personal care, apparel, automative, household items, packaged goods, health care and travel/financial services. In the public access videotex market, the first generation of systems relied on a network architecture. This meant dumb terminals hooked through phone lines to a central computer. There was a central on-line data base. The average display times in these systems were too slow—four to six seconds in many of the early systems. They had cartoon-quality graphics in fairly large data bases because it is expensive to work with advertisers to change the data bases.

This market will probably move in the direction of PC-based off-line terminals rather than a network architecture. Data bases will be stored locally, not centrally, on hard discs that operate close to the terminal itself. Display time will be reduced to one or two seconds, to keep the customer interested. A good portion of the frames in the data base will be photo-quality images, and bases should be smaller, more focussed, and more dynamic.

In the home market the first generation of systems emphasised graphics as well as text, used a videotex terminal rather than a PC, and was built around local delivery systems, with information about a specific market. We are now seeing the development of the second generation of home videotex systems which are moving away from colour graphics to speed the displays, and emphasise text and information content rather than visual images.

There will probably be a third generation, however, that will add photo-quality images to the text and graphics systems that predominate today. The systems will incorporate cable television and phone lines to provide two-way television services (images sent to the home via cable; data sent back to the central office via phone lines). An important ramification of such a set-up will be that a lot of the hardware necessary to support the system will not be located in the home. It can be housed in a data centre, or located on telephone poles where that investment can be capitalised and controlled by the system operator. Ultimately, we will see a fourth generation, a fully integrated system incorporating television, video-on-demand, and videotext. That would seem to be what the consumer wants—a simple, easy-to-use system that provides easy movement between interactive video broadcast and text-based delivery system.

The economics

Economics are a major reason for the continued growth of electronic shopping. The economic gains can be illustrated through two case studies.

First, profitability of a department store will be compared with that of a mail order business, and then show the gains to be made through electronic shopping. The profitability of a department store is basically an industry-average profit and loss statement for a large department store: 42 per cent gross margin, relatively small advertising budget, and relatively high operating cost. Carrying hundreds of thousands of items, the department store has high inventory costs. Margins are high initially but because of competition there is much price-cutting towards the end of each season. Operations are complex, with a distribution centre, central office operations, and large store with high labour costs. The net operating income (pre-tax) is about six per cent.

By contrast, the mail order business has fewer items in inventory, and a more unique assortment that is not as subject to competitive pressure. Mail order gross margins are higher, on average, than retail margins; 45 per cent has been assumed in this example. The advertising costs for the mail order house, however, are high because the business runs on media and direct mail promotions. That figure is set at about 20 per cent. Operating expenses are lower as a percentage of sales, however, because of the efficiencies that can be achieved in order-processing and fulfilment systems controlled from a central warehouse. The net margin (pre-tax) for industry is about 10 per cent.

Despite this kind of success, mail order businesses have historically been limited by the total number of mail order buyers in the United States. There are only 50 mail order companies with sales over $100 million, for example, in contrast to more than 1,000 conventional retail companies of that size.

This is where electronic shopping makes its presence felt. Electronic shopping offers the mail order retailer the potential to support significantly larger operations than have traditionally been achieved. New customers can be reached through terminal-based systems instead of direct mail, and, over time, a new group of non-store shoppers can be developed. In addition, the financial characteristics of a terminal-based business can be an attractive option for the marketing department. Instead of emphasising variable costs, such as mailings, the electronic merchandiser faces costs that are largely fixed, providing greater upscale potential.

A second case study showing the advantage of electronic shopping is the supermarket, illustrated by the profitability of the conventional industry-average supermarket, with a 22 per cent gross margin and a net margin of less than one per cent. The pressure on these margins is increasing as the competition from warehouse stores builds. As a mature industry, supermarkets are generally well managed, so there is not too much room for cost reductions. The electronic supermarket, on the other hand, can function more efficiently. It can work in a number of ways. The simplest is to print up a product listing catalogue from which customers can pick out their orders and call them in. Orders can then be delivered or picked up. Later, this ordering process can be automated, creating a fully electronic shopping experience.

The savings become significant for the retailer because instead of a chain of large stores, the retailer has one or more warehouses or distribution centres. These are highly automated and systematised because the customer does not have to interact with the merchandise. Costs can also be cut in energy and other key areas. The result is a very dramatic difference in pre-tax operating profit—from 0.6 per cent of sales in the conventional supermarket to 5.6 per cent in the electronic supermarket. Electronic shopping is important not because it can make use of the latest technology, but rather because it changes the fundamental economics of retailing. Established retailers must become aware and take advantage of these developments in order to maintain growth and profitability through the 1990s.

10

Privacy and security on videotex systems

Hermann Maurer

Main issues of privacy and security

This contribution starts by presenting a list of problem areas concerning privacy and security. The issues raised are discussed, some solutions are given and, finally, attention is focussed on an aspect which has been neglected for so long. Yet with improvements a number of potential abuses can be prevented; this relates to identification-procedures and page charges.

The problem areas usually mentioned in connection with privacy and security in videotex-systems can roughly be classified as follows:

Loss of information: How can information-providers protect themselves against inadvertent destruction of valuable information (due to system failure)?

Destruction of information: How can information-providers protect themselves against intentional destruction of valuable information by an 'enemy'?

Privacy of information: How can information-providers be sure that no unauthorised personnel has access to certain critical information?

Privacy of electronic mail: How can the sender of electronic mail be sure that the messages are only read by the intended receiver?

Information piracy: How can information-providers protect themselves against piracy of information (for example, the copying of a piece of telesoftware onto a floppy disc to avoid page charges the next time the software is needed)?

Junk mail: How can users make sure not to obtain large volumes of undesirable mail?

Security of banking: How can users be sure that their bank-account is not manipulated by unauthorised persons via videotex?

System breakdown: What can be done to make sure that vandals do not cause system breakdown by editing or mail overload?

System dependency: What can be done to ease the problems of a major system failure of multi-day duration at a future time when a good deal of every-day life is transacted via videotex?

Big-brother syndrome: How can users be sure that somebody is not keeping records of what pages they retrieved when and how often (compiling a profile potentially embarassing to the user)?

Page-charge frauds: How can users be sure that they are not billed for charges of pages they never retrieved (due to a system error or deliberate fraud by some other user)?

Each of these problem areas is discussed in greater detail below.

Loss of information

The loss of a serious amount of information due to failure of the videotex-system is highly unlikely, due to sophisticated back-up procedures. Still, with vital information some back-up at the information provider's end (on floppies as new data is generated) is certainly recommended and much practised.

Destruction of information

The intentional destruction of someone else's information requires access to the necessary passwords (or the cooperation of personnel operating the system). This issue will be explored further under identification procedures.

Privacy of information

In most videotex-systems information can be made accessible for the public, for members of a closed user group or for no one but the information provider. Thus, non-authorised access to information is possible either by acquiring the proper access codes (this problem is dealt with further under identification procedures) or for the system-operator. In contrast to the case of destruction of information the system operator and his employers are less above suspicion when privacy of information is concerned: nobody in his right mind would send plain-text telegrams on very secret issues, and nobody in his right mind would store information of high secrecy value in plain-text in a videotex system. However, this does not mean that sensitive information cannot be stored in videotex, it certainly can, but it should be stored in encrypted form, only decryptable if the proper key is known. More specifically, highly private information should be stored using one of the many cryptographic schemes available where the decryption key is only known to authorised persons. The encryption/decryption process can be carried out without much user intervention by the user's terminal if an intelligent terminal is used.[1,2,3] (This is just one more argument for why dumb-videotex terminals should be replaced by videotex-micros rapidly.[1] Also, Smart Cards can be used for the encryption/decryption process).

Privacy of electronic mail

Much of what has been said for privacy of information is applicable here, too. In particular, highly sensitive material should be sent only in encrypted form—encryption and decryption carried out by the user's intelligent terminal. It is surprising that this kind of application of the intelligent terminal has not attracted more attention. (More detailed accounts of encryption techniques combined with videotex are given in the references listed for this chapter at the end of this book.[3,4,5])

Information piracy

This problem becomes particularly pressing when offering valuable telesoftware via videotex: the idea of offering software in this fashion is that the charge per usage can be kept small, since the user will retrieve (and hence pay the charge) a large number of times. This scheme fails, if the user is able to copy the software in some fashion on a local storage unit (floppy disk) and use it from there whenever again needed. Still worse, the user may even decide to pass on the pirated software to friends. One way to solve the problem is to introduce special hardware.[6] A simpler, if less secure approach, is taken in safeguarding MUPID-telesoftware in Austria, Germany and Switzerland. The basic idea is that the software to be protected only works if the connection to videotex exists and if a certain page (the one with the page charge) is retrieved from the videotex system. It is worth noting that copying that page onto a local storage medium does not help since the program checks whether the information is received from videotex. Thus, while users can copy the software on local storage if they have such available (and are encouraged to do so to save down-loading time) the program does not start unless the page with the charge is retrieved. To beat this scheme the user would have to analyse the object-code responsible for above procedure: specialists will certainly be able to do so if they have enough patience but this is acceptable. After all, what has to be prevented is easy copying, not a one-time breaking of the mechanism by specialised hackers.

Junk mail

It can be a serious problem if users have to wade through dozens of unwanted and unsolicited electronic letters (it takes longer to throw them away than paper junk mail, and costs telephone-connect time!) but it is a problem of privacy and security only in a very peripheral sense. Anyway, while some videotex systems allow the user to define a 'black list' (no mail from persons on this list) or a "white list" (only mail from persons on this list) it is our belief that such issues are best resolved by using an intelligent terminal which can use fancy software to weed out whatever messages are deemed undesirable. (Observe, in passing, that such software usually requires some individualised information (that is, a list of numbers) which can be stored on a local storage or in battery-buffered memory, if available, but else has to be stored in the videotex system. Hence, for applications such as this users who are not information providers need pages in the videotex-system to store information on. Such pages are, unfortunately, not visually provided by the main videotex system but only on third-party computers.

Security of banking

This is clearly one of the most sensitive issues concerning security in videotex systems. It is also the one most discussed, the one with the largest number of incorrect new reports and probably one of the most secure aspects of videotex. Indeed, to date no single case of anyone breaking into a videotex bank account has occurred to my knowledge. A case such as the Chaos-Computer-Club scandal in Hamburg, which was widely reported as a case of cracking a videotex bank account, was of a completely different nature: it was a page-charge fraud and is discussed below.

The usual techniques employed for the security of videotex-banking (such as the personal identification number PIN, internal level of identification, and transaction number TAN, a separate number for each transaction, each number only usable once) assure a fairly high level of safety, certainly higher than the use of credit cards or the like. To reassure the public and to protect the consumer I feel banks should agree to one additional safeguard: a customer should be able to specify for each of his or her accounts an upper limit to the amount allowed to be transfered from the account via videotex per day and per month, a limit beyond which the bank is responsible for money misused. For instance, I would probably choose values such as DM 2,000 per day and DM 10,000 per month, assuring me a reasonable disposability yet limiting my losses in case of a real mistake (such as loosing all my passwords, and the loss of TAN's and not noticing it). With such limits, banking via videotex is certainly safe with the safety features already making banking this way rather tiresome. To both improve security and convenience, Smart Cards are considered the best alternative but require extra hardware.

System breakdown

There is no absolute guarantee against system breakdowns, both caused by accidents or by deliberate vandalism. However, the consequences can be minimised by various measures such as good back-up procedures, automatic restarts, and back-up systems (particularly by load-distribution to other nodes of the videotex-network if some fail).

Another interesting point is that the ironic saying 'money is educational' also works in videotex: much abuse and vandalism occurs because services are free or not expensive enough. Typically, system overload by too much electronic-mail in the German videotex system was mainly caused by indiscriminate electronic distribution of low-value information possible only since no charges were levied (at that point) for messages: computer vandals distributing thousands of pieces of electronic mail by means of teleprograms, or causing the editor of the videotex-system to blow up due to continuous log-on/log-off sequences have ceased their activities abruptly once moderate charges were collected for each piece of mail and each log-on of the editor.

System dependency

The danger that our society (or a part of it) becomes dependent on videotex (once videotex is widespread) and hence becomes totally paralysed in a very serious way should a major system failure occur (due to an earthquake, a fire or terrorist-bombing) is considered by many a purely hypothetical and philosophical problem, not one to be worried about at this point. I would like to disagree with such an attitude: early attempts have to be made to minimise the probability of total breakdowns, much in the sense as is done today with electric-power supplies: even the loss of a large number of power-plants will not cause a complete blackout due to more and more improved electrical distribution and switching methods (accelerated by lessons learnt from the famous major blackouts in the New York region in the 1960s). In a similar way, a videotex network should be designed so that it continues to function (possibly at reduced performance) as long as a single videotex-computer in the network is operable. The architecture of some systems which depend for crucial functions on the availability of one super-control-system, or which allow the user only access to one particular node have to be viewed with extreme suspicion: the first serious crisis after videotex has taken over important functions (as it will), will amply show that such a concept is irresponsible. The public is fully justified in exerting pressure on operators of videotex systems to make their systems at least as resistant against failure of some of the videotex computers as electrical networks are nowadays against the failure of some of the power-plants. This is technically achievable and should have high priority.

Big-brother syndrome

It is often claimed that videotex increases the possibility of supervising and controlling people in the sense of George Orwell's Big Brother in *1984*. The argument usually is that users, as they log-on, have to identify themselves (in order that fees can be charged to their accounts); in this fashion, the videotex-network (or third party computers) could be programmed in principle to retain for each user which information has been viewed, how often, how long, at what time and (possibly even) at what location. Thus a 'profile' could be compiled showing such things as political affiliation, sexual preferences and a variety of other habits, thereby providing information which could potentially be used against the videotex customer at some stage. Such arguments should not be taken lightly at all. More people have things to hide than is superficially accepted; statements such as 'those who live properly have nothing to fear' are nonsense, uttered only be fools or hypocrits. For instance, 23 per cent of use will die of cancer—when some of us find out and search the medical pages of videotex desperately for some help we are unlikely to want the videotex operator (or anyone else) to know.

Hence, our information profile must remain a secret to everyone. Those who don't understand this yet are urged to consult further literature.[7] The problem is reduced by the data-privacy and protection laws available at various levels depending on the country and by the fact that videotex-operators usually have better things to do than to surveil their customers; also, they are usually subject to some regulation requiring to keep information as described above to themselves, anyway. Nevertheless, rules and laws are rarely a genuine answer. What is much preferable is a system which makes the kind of surveillance mentioned as hard as possible. This has led to the introduction of the notion of anonymous access and anonymous charging.[8] Under such a scheme videotex-users have a modem with no individualised identification and can use the videotex system by just dialling up the system without need for identifying themselves. Such schemes are clearly feasible if no time charges are collected for the use of videotex (as is the case in many countries such as Germany, Switzerland and Austria) and if no frames with page charges are retrieved. For frames with page charges anonymous accounts can be used: a customer, without identification can buy (at banks or post-offices) a password worth, say, DM 50 and log onto the system using that password (which cannot be traced). All fees are charged against the password until its value (in this example DM 50) is exhausted, at which point a new password is used. As incredible as it may sound, it is even possible to order merchandise and get it delivered without ever being identified.[8]

Anyway, the point is that by minute changes in the charging mechanism of videotex-systems (against anonymous prepaid passwords rather than against identifiable customers) surveillance of videotex-users can be made close to impossible (unless phone calls are tracked systematically, a task not easy in our current phone-network).

Hence this much anonymity should be introduced in all videotex-systems: it is irresponsible not to do so. As explained later, a host of other problems are solved with this approach at the same time. As with intelligent decoders and telesoftware[1] Austria was the first country to introduce anonymous videotex access: it worked as predicted. In the meantime anonymous videotex access is also available in other countries, such as Switzerland.

A final word concerning an individual identification sent out by modems is necessary: it does help to identify videotex-customers hence may increase security. It certainly reduces privacy (i.e. increases surveillability) dramatically. This is a typical instance where the aims of privacy and security are somewhat at odds with each other. However, since reasonable methods for security exist (see below) modems with individualised identification should not be used to any great extent.

Page-charge frauds

One of the most celebrated cases supposed to demonstrate the lack of security of videotex systems was perpetrated by the Chaos Computer Club (CCC), Hamburg. CCC had found out (during an exhibition) the identification of another videotex user X. CCC proceeded as follows: CCC edited a frame under its own password with a frame-charge of some DM 10. Then, CCC dialed up the videotex system using X's password and retrieved the frame at issue 15,000 times using a teleprogram . . . amounting to transfering a page charge of DM 150,000 from X to CCC. CCC, bent on obtaining negative publicity for videotex, chose X to be a bank so that the press would hopefully report (incorrectly) that CCC had broken into a bank via videotex. The press obligingly did so.

The lesson taught by CCC is an important one, but few people seem to have understood it. There is nothing easier than performing page-charge frauds in most current videotex-systems: you determine a user password by spying (or trial-and-error, see below) and you make sure that one of your frames carries a hefty page charge. Then you dial up using the illicitly acquired password and retrieve your own frame a few thousand times (or, if you want to go undiscovered, just a few times). Depending on the laws this may or may not be a criminal offence; if you are careful about it how can anyone prove it was you who retrieved your frame? What I am trying to say is this: page charge and identification procedures as they are commonly in use today are not very secure. Some changes are urgently needed.

What is wrong with current page charge and identification procedures?

The point most basically wrong with page charges is that they are potentially unlimited. If you loose your wallet, you loose as much money as you have in it; if you loose your card with which you can cash money at one of those automated tellers you loose at most up to a certain amount a day (if someone manages at all to misuse it); if you loose your videotex identification and password someone may use it to collect a few hundred thousand DMs in page charges from you within a few hours. (It is true that when you go to court on this you may eventually not have to pay—but you may have to win an interesting legal battle, first). One step against such blatant misuse of page charges is to impose a certain limit per day or per month on page charges, the amount may be decided by the customer when first registering for videotex. (To my knowledge this very applaudable approach has only been taken by the Netherlands, so far).

An upper limit on page-charges certainly helps, but more subtle ways of criminally using page charges are possible which are not prevented this way and will easily go unnoticed: suppose a user finds the access number of n companies and uses them to collect (fraudulently) DM 100 a month in page charges from each of the companies. The amount at issue will go unnoticed because of the large phone-bills of the companies involved. And if n is sufficiently large, say $n = 50$, some DM 5,000 of page charges can be collected month by month. This type of fraud depends on how easy it is to determine access codes of other videotex users by trial and error. Unfortunately this is comparatively easy: standard videotex-systems have an identification code of at most nine significant digits, followed typically by four separate letters of a personal password which can be changed by the user. Assuming a system has 10,000 users, by trying all possible 9 digit combinations one by one a valid user identification is found (using a teleprogram without human intervention) every $10^9/10^4 = 10^5$ tries. Assuming 300 tries an hour it will take about two weeks to find an identification and somewhat longer to determine the proper password. Thus, within a few months (break-in attempts in some systems were already successful after a few days in some of our experiments) all necessary identification data of some other user can be determined, a moderately alarming fact.

An easy way to dramatically increase the effort required to 'break' a system is to use longer identification or (as practised in Austria) to combine identification and password into one string (increasing the average break-in time to a few years), or even to use individual modem identification (not recommended for privacy reasons and for sharply reducing flexibility when travelling).

In summary two points emerge. Firstly, rather than 'no limit page-charges', page-charges with monthly upper limits or, better still, anonymous page-charges are recommended. Anonymous page-charges have two further important advantages: they preserve privacy and users can be allowed to access videotex (for example in a company), each with a separate anonymous account and hence with precisely allocatable pages charges per person. Secondly, the current uniform identification mechanism (lumping simple videotex-use, electronic mail, page-charges and ordering all together) is highly undesirable. Rather, a non-uniform identification scheme, outlined below, should be considered.

No identification required for access of ordinary videotex frames carrying no charge (anonymous access)

Identification of an anonymous prepaid account required to pay for page-charges or the like (anonymous charging)

Identification of an anonymous P O Box required for retrieving electronic mail from such a box (a user may well decide to have an anonymous electronic P O Box only, or in addition to his public one)

Identification of user for sending a message or ordering merchandise (in this instance anonymity could also be preserved but is probably not desirable

Identification for editing pages

Identification for carrying out financial transactions

Identification for changing any of above identifications

The outlined 'layered' approach to security corresponds more to reality than the usual uniform approach and is implemented readily: it is hoped that such implementation will spread rapidly.

11

Audiotex: the telephone media

David Shorrock

What is audiotex?

Audiotex is the audio equivalent of videotex, but instead of a television terminal and pictures, it uses the humble telephone. Behind audiotex is software that allows information from an on-line database to be passed to a voice-mail system, where it is then interpreted and delivered to the user over the telephone as a natural, spoken message. The voice-mail system is a computer that stores individual words, word groups, part or whole sentences, or complete messages, in digital form. When required, these are matched by the voice-mail computer to the information coming from the database, combined, and then relayed over the telephone.

Audiotex: how it works

The heart of any audiotex system is a voice-mail processor. Figure 11.1 shows a schematic of a typical processor. The line cards connect the processor to the incoming telephone lines. They contain microprocessor based circuitry to detect the tones from the callers telephone, as well as the codecs that digitise the incoming speech. These same codes work in the opposite direction, recreating analogue speech from the stored segments of digitised speech.

The line cards are connected by means of a bus with the central processing unit (CPU) of the voice processor. The CPU performs all the control functions of the voice processor including the addressing, storing and forwarding of messages.

When a message is recorded on the system, the CPU gives each message a unique index number and stores it on disc. That message, be it a single word, partial or whole sentence can then be recovered subsequently by that same index number. In this manner, an entire vocabulary of words and phrases can be built up, in one or more languages.

So, far, it has been assumed that the audiotex system is based around voice-mail hardware, in which natural speech is stored in a vocabulary and data from an external host is used to construct phrases and sentences. However, systems are available that use text-to-speech synthesisers which, although sounding computer generated, have the advantage of being able to convert virtually any text to speech.

The voice-mail processor, despite having the processing capability of a mid-range minicomputer, is generally not used to maintain the database for the given application. Often this is because there is an existing database on a host, which is already serving an existing terminal network. Hence the audiotex system is an extension to that network, with the voice processor accessing the host in the same way as a terminal.

The manner in which the voice processor and the database host are configured is usually dependent upon the rate of change of the information. There are five common configurations. *Duplicated database:* if the data is not changed by the audiotex application, then the relevant data can be transferred from

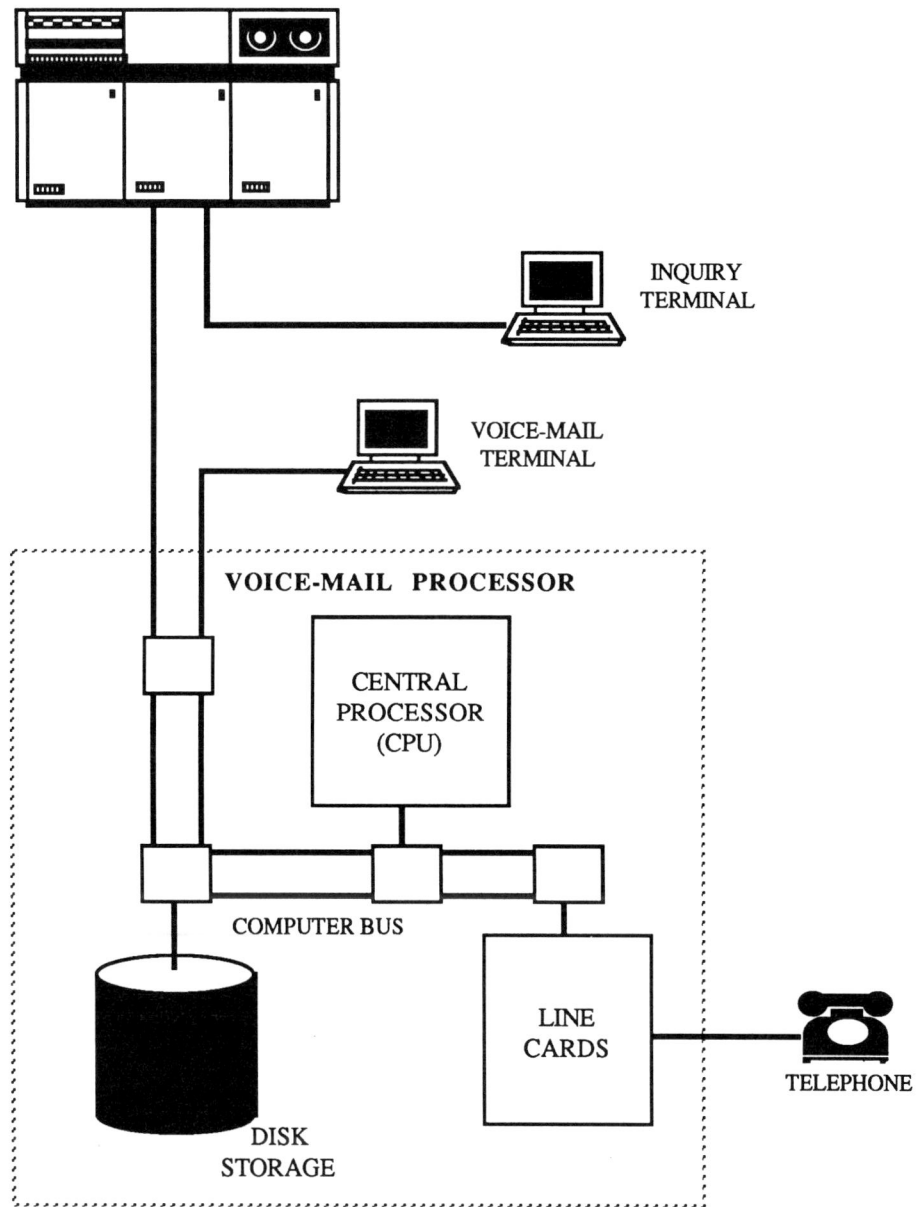

DATABASE HOST

INQUIRY
TERMINAL

VOICE-MAIL
TERMINAL

VOICE-MAIL PROCESSOR

CENTRAL
PROCESSOR
(CPU)

COMPUTER BUS

LINE
CARDS

TELEPHONE

DISK
STORAGE

Audiotex System Schematic

Figure 11.1

the host to the voice-mail processor, where it is stored. *Duplicated database with automatic data capture:* if the host database is changing rapidly, then the database can be duplicated as above, but with the changes, whenever they occur, also being fed to the voice-mail processor. For example, this approach is often used on systems that are providing share price information. *Duplicated database with timed capture:* if information from the entire database is regularly requested, but the changes are relatively infrequent, then the preferred approach is to maintain separate databases, with the voice processor database being updated periodically from the host. *Selected data with timed capture:* if the enquiries are such that there are a large number of requests for a restricted set of information, which changes infrequently, then the processing load can be reduced if only that set of data is updated periodically on the voice processor. For example, consider the case of a commodity price system that frequently receives requests for the price of gold on the London market. The gold price changes only twice a day, and is otherwise fixed. Hence the voice processor can be programmed to request the new price from the host at two set times a day. At all other times, the processor can relay the price without interacting with the host. *On-demand inquiry:* if the database is changing frequently, but the requests are relatively infrequent, then the preferred relationship is for the voice processor to access the host for each enquiry. This approach is also used when the audiotex system updates the database, as in the case of a retailing system.

Audiotex: the human interface

The comparatively limited communication channel offered by audiotex has led to a careful study of the processes of user interaction. This has shown that the most effective applications are those in which the user is first told to make a choice, the choices are then listed, each choice is then explained, and then finally, the user is requested to make a decision. In order not to frustrate frequent users of the service who no longer require this 'audio menu' and who wish to make an immediate choice, the system should be capable of being interrupted by the user and a choice made at any time. For instance, a secretary who regularly books business flights on an airline reservation system, and who knows the city codes and flight times, can go straight to the booking process without listening to a lengthy preamble.

The number of choices that a user is allowed must be restricted because people tend to forget long lists and only remember the first and last items. Experience has shown that the practical limit is six choices before the user becomes confused, and that three is the optimum.

Audiotex applications: an overview

Audiotex has found wider use in the USA than in Europe. American users include: airlines, financial information service providers, retailers and the medical profession. As described elsewhere in this section, Trans World Airlines operate a number of audiotex systems including a crew scheduling system that handles several thousand enquiries a day.

Flight schedules and updated aircraft arrival and departure details are made available to the public by Pacific Southwest Airlines audiotex system. Using their touch-tone (or multi-frequency) telephones, passengers answer a series of verbal prompts as to their departure and destination. The system then responds with the appropriate route and flight times.

In the financial services sector, audiotex services compete with videotex in providing the latest information on share prices, foreign exchange and currency movements. In the US, the Dowphone service has been available for a number of years, and in the UK, British Telecom have relaunched their Citycall share price information service to provide a similar service.

The *Physician's Desk Reference* is a standard work used by general practitioners in the USA. This loose-leaf bound volume is published annually, with monthly updates. The publishers are now offering an audiotex update service so that, for instance, a doctor wishing to check on drug interactions, can dial the service and then key in the number of the drug in question. The service will then read the relevant information to him.

Retailers are using audiotex to advertise goods and services. A caller to an audiotex mail-order service, will listen to the descriptions of the items, and then can order by keying the item number, the required quantity, and his credit card number. If the credit card is valid, the system then requests the caller's name and the address to which the goods are to be sent. This information is recorded for a human operator to add manually to the other information which the audiotex system has passed directly to the computer.

In June 1987, the Trustee Savings Bank (TSB), launched an audiotex service for their personal customers. This is in contrast to other UK banks and building societies that have offered videotex services for home banking. TSB offer the rationale that the UK has some 28 million telephones, whereas there are only 100,000 videotex terminals, few of which are in the home. The TSB Speedlink service allows customers to check the balance of their accounts, transfer funds between their accounts (for instance from a deposit to a current account) and to pay prearranged bills such as gas, electricity and credit cards. As the majority of domestic telephones in the UK use pulse as opposed to multi-frequency dialling, for a payment of £10 TSB provide their customers with a calculator-sized keypad tone generator which is held against the mouthpiece. TSB's Speedlink is the first large scale public audiotex system to be implemented in the UK, although other smaller systems have been implemented by companies for business-to-business communications.

Audiotex futures

Dial-out

If in the past, audiotex has been a passive media, then in the future, it will become proactive. Audiotex systems are now available that dial out. Voice-mail messages can be delivered automatically to a potentially unlimited number of telephone subscribers.

One of the first applications of dial-out audiotex systems has been telemarketing, with retailers canvassing large numbers of consumers via their home telephones. In the US, some states have bowed to pressure from consumer groups and such services are prohibited by law. In Europe, dial-out audiotex is still in its infancy and such restrictions have yet to be put in place.

Risk Arbitrage Monitor Inc, a firm of Wall Street investment advisers, are professional users of dial-out audiotex. Their service dials out to clients, not by telephone, but by radio pager, to notify them of information that they have requested. For example, to alert clients of a change in the investment position of a particular company, the message is first stored in a voice mailbox; the relevant subscribers are then notified by the paging system, with the box number of the message; the subscriber, who receives that box number on his pager can then phone the number at his leisure. The system is secure in that the message will only be relayed to the subscriber if he correctly enters his seven-digit password when he calls the service. This dial-out service allows any number of subscribers to be simultaneously alerted to information, in any of 35 cities across the USA.

The future of audiotex extends beyond the simple relaying of information. By using the 'if-then' capabilities of computers, future audiotex products will allow users to specify conditional responses such as "if my share prices rise above £5 per share, then page me". More complex tests might be "if the composite of my portfolio moves more than three points, call me". The investor no longer needs to track the market, the audiotex system will do it for him. If any of his shares move significantly in either direction, the system will page him, wherever he might be.

Voice recognition

Many market observers believe that audiotex will become more widespread when voice recognition becomes a reality and users can move away from the need to use a multifrequency telephone or carry a tone generator. In December 1987, the US company Datapoint demonstrated a simple voice recognition system capable of recognising the numbers *zero* to *nine* and the words *yes* and *no*. Unlike many previous systems which are user-specific and require the user to 'train' the system by repeating the same word a number of times, the Datapoint system is speaker-independent. The benefit of such technology is that the system appears more natural as the user just speaks the words instead of having to key a number.

Another major player in developing voice recognition systems is British Telecome. BT are currently running trials of Voice Operated Database Inquiry System, VODIS, which is being developed under the auspices of the UK's Alvey project, with a number of participants including Cambridge University. The purpose of the work is to develop a system capable of recognising continuous speech, as opposed to the one-word responses that the Datapoint system is capable of recognising. The VODIS demonstrator is a mock train timetable enquiry system for the route between London and Aberdeen. Despite promising results from the human factors trials that have been conducted with VODIS, British Telecom estimate that it will be the early 1990s before such systems become commercially available.

Pessimists however, argue that even when voice recognition technology becomes a commercial reality, it will face consumer resistance because most people feel uncomfortable when talking to machines.

12

The audiotex industry and markets

C William Reed & Bruce Kushnik

It has become almost commonplace in our era to refer to the 'birth of an industry', simply because we have witnessed so many: computers, microchips, artificial intelligence, video games. However, it is rarer to witness the birth of an industry that has yet to become aware of itself as an industry. Yet that is precisely the position in which the interactive voice, or as LINK and National TeleVoice (NTV) prefer, the interactive telephony industry finds itself.

If telephony is the art, science and practice of using the telephone, then interactive telephony is the process by which a human interacts with a computer or a tape system over the telephone network. The input device can be a touchtone telephone, or in an increasing number or cases, some form of speech recognition technology.

Evolution of the industry

In its broadcast sense, the interactive telephony industry was born when the first person, probably the first switchboard operator, was asked to take a message and relay it to another person. By the 1920s the telephone had become accepted as a necessary household appliance, and interactive telephony progressed to its second stage—answering services.

The first level of automation was introduced by New Jersey Bell in 1927: by dialing a certain number, the caller could hear a recorded voice tell the correct time. Although additional services were added through the next four decades (dial-a-prayer, weather information, recorded messages from the phone network and widespread acceptance of answering services), the industry remained in an embryonic phase because of a lack of the technology required to process the voice signals electronically. In addition, regulatory statutes forbid any equipment not sold by the telephone company to be attached to the telephone network.

By the mid 1970s the pieces began to come together. Microchips had revolutionised the computer industry, and began to bring the cost and size of devices into the range of the average consumer. The personal computer was introduced in the late 1970s, affording inexpensive and available processing power. And voice messaging, the first entry into a truly automated era of interactive telephony, was introduced.

The first voice messaging system was marketed by Sudbury Systems in 1974, and was built around their popular dictation equipment. It was not until 1980 that Gordon Matthews, founder of ECS Telecommunications Voice Message Exchange, invented and patented the first digital voice messaging system. The company he founded, now known as VMX, is still one of the major players in the industry.

Emergence of complementary technologies

Around the same time, several other technologies and applications began to spring up that were based on an ability to process the human voice:

in 1971, Threshold Technologies delivered the first commercial voice recognition equipment (speech-to-text) to Federal Express.

consumer answering machines became widely available in the late 1970s, based around audio tape technology.

the first commercial text-to-speech products were developed and marketed in 1976.

auto-dialed recorded message players (ADRMPS, pronounced Ad Rumps) began to be used by telemarketing companies. These machines are capable of dialing thousands of numbers and delivering a recorded message automatically.

touchtone telephones, based on dual-tone modulation frequency (DTMF) technology, became widespread, and are now found in over 50 per cent of US households. These phones enable the caller to input data via the tones on the keypad to a distant computer.

private branch exchanges (PBXs) replaced the local central telephone company office for many businesses, and provided an on-premise switching system.

automatic call distributors (ACDs), were introduced in companies receiving thousands of incoming calls. Now rather than a busy signal, hundreds of people could be put on hold simultaneously while waiting for a live operator.

While all these innovations in voice technology were becoming available, data processing exploded and another entire industry was born around the ability to manipulate and communicate alpha-numeric data. The technology required to manipulate data, which is naturally digital, is much less complex than the technology required to process or produce a human-sounding voice, which is naturally analogue. Consequently, many of the major providers of information services—newspapers, banks, computer companies, and even the telephone companies—began to experiment with methods of delivering data to homes and businesses over the quickly growing base of data terminals. AT & T not only spent millions on trials of videotex systems, but it was so interested in getting into the computer business that it built the entire structure of the divestiture decision around its ability to manufacture and sell computers, and eventually information services.

Yet no matter how simple the technology for delivering data becomes, there is still the problem of the last inch—the human interface. This is precisely the problem that the interactive telephony industry resolves, and it does so firstly by making the telephone keypad into a ubiquitous data entry and data access terminal and, secondly, by making the human voice, rather than the fingertips, an input/output vehicle.

These two factors provide the unifying theme around which a variety of new industries have arisen. The only problem is that all of these individual industries have yet to recognise the blood ties between the siblings in this new and constantly growing family.

Five basic technologies behind interactive telephony

Until 1984, the technologies involved in interactive telephony could be relatively easily separated into five categories. Since that time, these categories have begun to coalesce into a continuum of hardware products that are differentiated almost entirely by software applications: consequently LINK/NTV now believes that they are beginning to form a larger whole which we have named interactive telephony. The five initial categories are described below.

Interactive voice response systems

These systems were initially created peripherals to mainframe computers enabling them to communicate with users by means of a digitised voice. They are used primarily for field service, order entry and data collection applications.

Audiotex systems

There are two distinct types of audiotex hardware systems. *Announcement Systems*, now called digital recorders, began as glorified telephone answering machines. They are accessible from any telephone, since they do not have touchtone interactive capability. These systems are primarily used by non-interactive *976* systems, whose name derives from the first three numbers the caller dials into the special telco network designed to handle these announcements. *Interactive Audiotex Systems* are distinguished from the simple announcement systems by their interactive capability. These systems utilise touchtone telephone input to access and interact with a recorded message.

Voice messaging systems

These grew out of the demand for answering services and tape-based answering machines, but really became an industry with the introduction of digitised voice signals. They are built around a computer system capable of storing and forwarding recorded messages controlled by touchtone telephone input.

Interactive voice/data terminals

IVDTs, which are basically personal computers enhanced with a telephone handset and communications capabilities, a modem). Their functionality depends on the ability of the network to transmit voice and data signals over the same line. Many of them have added text-to-speech and voice messaging capabilities, and consequently have entered interactive telephony somewhat obliquely.

Autodialers or telecomputers with audio

ADRMPS are used by telemarketers to replace or supplement live operators. A computer dials thousands of numbers, and plays a recorded message to the person who answers the phone. These machines have added touchtone recognition, which enables the computer to enter the realm of market research by asking questions which the person can answer via touchtone input.

Voice messaging

Of these five types of equipment, voice messaging has been by far the most successful and the most talked-about to date, for three primary reasons. Firstly, the telephone tag is perceived as a major annoyance by most companies, and voice messaging provided a natural solution; secondly, PBX vendors realised that by offering voice messaging add-ons they could not only increase revenues from the installed base of PBXs, but they could also market a new feature to prospective customers; finally, the costs began to drop dramatically after 1983: from $250,000–$300,000 for a 1,000-user system to between $15,000 and $25,000 for introductory systems.

Nevertheless, voice messaging still has not lived up to the exorbitant expectations people placed on it when it was first introduced. Part of the problem has been selling a feature whose cost justification is not immediately apparent. In the same way that electronic mail has had a slow beginning, voice mail systems need a 'critical mass' of users before the system becomes truly efficient. But to the manager in charge of paying for the system, it remains an expense item that can be put off until next year's budget.

In addition, LINK/NTV believes that voice messaging as such will be surpassed in the marketplace very quickly by multi-functional machines offering the capabilities of all the equipment categories mentioned above.

Interactive voice/data terminals

Perhaps the least successful technology in interactive telephony has been the interactive voice/data terminal. Again, observers were wildly enthusiastic about the devices when they were first introduced, but the same technology which produced them may also have defeated them, at least for the time being. By the time they were available in any quantity and at a reasonable price, personal computers had proliferated to over half the desks IVDTs were supposed to penetrate. Suddenly an attractive technology became a second generation or replacement technology. In addition, the machines were designed to be used by upper-level executives who did not want to be responsible for the communications link available; they all had secretaries to handle the telephones for them.

Moreover, many companies (LINK among them) remain dubious about the time-frame in which the integrated services digital network (ISDN) will be introduced. ISDN will provide an environment in which voice and data and even video can be transmitted over the same line simultaneously; however, LINK does not expect ISDN to be widely available until well into the 1990s, and even then it will probably be priced as a premium service.

Automatic dialers

Automatic dialers with audio capabilities have found a natural market niche in telemarketing. Not only does this technology enable a service bureau to automate its operation and greatly increase its efficiency, it also permits smaller companies to conduct their own telemarketing effort at very little cost. The machines are really only glorified cassette machines with autodial and message taking capabilities. Power-dialers, another class of automatic dialers, perform the same automatic dialing function, but also add an interface with a live operator, who then connects the person to the recorded message.

The potential of these machines is tremendous. Retail and mail order companies like Sears and Roebuck use these machines to notify customers that their orders are in; Shell Oil uses them to collect on its credit cards; the Democratic and Republican parties use them for polling and surveys; the IRS uses them to collect late payments.

In addition, some legislative quirks have opened new markets for this technology. Nuclear power plants are not allowed to begin operating until a satisfactory evacuation plan has been submitted, and at least one company has developed the software required for an ADRMP machine to contact the people responsible for executing the evacuation plan. In a completely different direction, many states now require schools to notify parents that their children are absent from school on a given day, and many school districts have discovered the benefits of automating that function. And of course nearly every business could use them for telemarketing applications to generate sales leads. However, legislative restrictions have had a negative impact on the use of autodialers in general because consumer groups argue that they do not want to receive unsolicited calls. Manufacturers and telemarketing companies are forming lobbying efforts in many states to combat prohibitive restrictions.

Interactive voice response

Perhaps the most versatile of the five technologies is the interactive voice response system, because it permits a caller to interact directly with a host computer through a telephone interface. The voice response aspect means that a person can obtain database information from any telephone through a recorded or synthesised voice. The interactive component means that a caller using a touchtone phone can also enter data or direct the computer to deliver a specific type of data over the telephone.

But the real versatility of an interactive voice response system comes from other, industry specific applications. Banks use the systems to enable customers to check account balances, see if a cheque has cleared, pay monthly bills, and transfer funds from one account to another. Brokerage firms can dispense stock quote information on a nearly-real-time basis. Airlines can provide customers with flight schedules, specific flight arrival information, and even provide complete ticketing services over the phone without ever speaking to a human operator. Service companies can provide customers with a locator service to direct the caller to the nearest branch office to the phone he is calling from. Almost all simple order-entry applications can be accomplished on an interactive voice response system. Retail companies can perform credit authorisation over the phone without a dedicated point-of-sale terminal.

The list goes on and on, and increases daily as companies begin to realise the benefits of selling creative applications rather than fancy technologies.

Market fragmentation

The interactive telephony industry is by no means a unified whole. Although to an external observer interactive telephony may be beginning to look like an industry, to the participants themselves this is not yet the case.

LINK/NTV has identified over 400 different manufacturers of equipment falling into one or more of the four categories listed above. For the most part the vendors have identified themselves entirely within a given category, and were more or less unaware of the existence of the others. Although they have developed functionally similar equipment, they market it to totally separate customer bases. An autodialer with audio capabilities may be advertised as a telemarketing system, while another type of autodialer, also with an audio functionality, is sold under the auspices of office automation. An interactive voice system may lack only the software and some memory capacity to make it into a voice messaging system, yet the manufacturer might be totally unaware of the potential of selling it as an office automation product.

Companies from different categories advertise in different publications, go to different industry shows and utilise different distribution channels. Users as well have bought into the rigid separation between the categories, and often do not realise the real capabilities of the machines they purchase. Nor do they realise they could have saved thousands of dollars per line simply by moving across to another equipment category. However, the advance of technology, combined with the slow growth of many of these individual markets, has produced not only a decided convergence within the capabilities of the hardware, but the initial awareness by manufacturers that the other markets exist and offer a tremendous potential.

Thus the interactive telephony industry resembles a family whose children were all adopted by foster parents and who never knew the other family members existed. LINK and NTV hope to provide the occasion, and the justification, for a great family reunion.

Factors slowing growth

Several factors have contributed to the slow initial growth of this new industry: lack of distribution channels, lack of applications marketing expertise, lack of software for applications and the cost of equipment.

Both the lack of established distribution channels and, among the manufacturers, a general lack of expertise in marketing applications for their equipment are significant contributors to the drag on industry growth. In 1986 LINK/NTV conducted an informal survey of 30 major interconnect companies, and nearly two-thirds of them did not carry voice messaging or interactive voice equipment at all, and the ones who did carry some type of equipment, usually a voice messaging system attached to a PBX, were marketing it as an add-on box, and not as a separate application.

Voice messaging is finally beginning to achieve acceptance in the marketplace, helped largely by the fact that its primary distribution channel is the PBX and ACD vendors marketing the devices along with their established equipment lines. Interactive voice response has not reached that threshold of acceptance, in large part because of a lack of awareness as to its potential applications. Another drag on the growth of this marketplace has been the lack of sophisticated software for application building. Up to this point, hardware vendors have been content to list a variety of applications on their brochures, but have given little support to companies interested in pursuing other applications specific to their situation.

Similarly, very few value-added resellers (VARs) understand all the potential applications for interactive voice equipment. Nevertheless, LINK/NTV feels that VARs will become a major force in marketing and distributing interactive voice equipment by focusing on applications, and that they will become a unifying force in the industry.

Admittedly, convincing a corporation of the benefits of either interactive voice or voice messaging equipment is not easy. The systems will be used by a large number of people, if not the entire company, and closing a deal requires a consensus of multiple departments. Even if the application is a familiar one, such as a sales support system for a large multi-regional company, several groups must be convinced of the benefits and then trained to use the system: the sales people themselves, the sales support staff, superiors wishing to get in touch with the sales staff, and probably the MIS staff as well.

Convergence creates an industry

Already the signs of convergence which will lead to a family reunion are becoming more evident. The number of new entries into the interactive market continues to grow. The hardware for several different applications is becoming similar enough for the software to define the functionality of the hardware. Voice messaging companies like VMX are starting to offer audiotex applications and will continue to enter the interactive market. OPCom, an interactive voice system manufacturer, now offers voice messaging as an added feature. Companies are beginning to market applications rather than hardware, emphasising software support and program development.

PBX and ACD manufacturers are beginning to see the benefits of offering some type of interactive voice capability. One of the primary marketing strategies of voice messaging equipment vendors has been to add a module on to an existing PBX system. Teknekron InfoSwitch, a large ACD manufacturer, now offers its voice response unit (VRU), AT & T's Conversant 1 system will interface with the AT & T PBXs. Other ACD and PBX manufacturers will be moving in this direction in order to increase their installed base of equipment. Consequently, small companies making standalone equipment may become more and more vulnerable.

Voice messaging and ACD functions are beginning to merge already. The addition of an auto-operator function to voice messaging systems makes it very similar to an ACD: it directs calls to extensions and can call forward, giving it a PBX-like function. Conversely, PBX manufacturers are adding voice prompts (for instance, auto-operator) to increase call handling capability. The voice messaging and ACD industries are currently separate, but are rapidly encroaching on each other's product lines; LINK/NTV expects that there will be a dramatic merger in the functionality of these two systems.

Comverse, with its Trilogue system, has targeted what it calls application controlled voice services. This area is produced by the convergence of person-to-person messaging (voice mail and electronic mail) and person-to-system messaging (voice response and audiotex). The Trilogue system offers standard voice mail, a dialog generator application for audiotex and auto-operator functions, and a text-mail integration feature.

Market trends

Excluding voice messaging, which is being sold as an office automation package usually in conjunction with a PBX, the only companies buying quantities of interactive telephony equipment right now are: banks, financial services companies and telemarketing companies. As applications software improves, LINK/NTV expects the market as a whole to benefit from the wider familiarity with the technology.

While all of the following trends are legitimate in their own right, they all must be interpreted within the context of the wider telecommunications environment with particular attention being paid to legislation.

The low end of the interactive voice market is developing rapidly, and will provide a major impetus to the market as a whole. These products offer some form of interface with a PC, and will bring interactive voice applications to companies that could not afford it previously. Systems cost approximately $1,000 a line, and can expand up to 64 lines.

The 'smorgasbord effect' is rapidly becoming the norm. As the similarity between types of hardware grows, manufacturers are offering machines that can do everything imaginable: voice messaging, speech-to-text, text-to-speech, voice recognition, voice-data inputs, data interfaces, host computer control, call detailing, audiotex, auto-operator, interface with PBX and ACD. The new IBM Communications Option is an example of this phenomenon at the low end, the AT & T Conversant 1 system at the high-end.

Prices are dropping rapidly, creating anomalies in the market and leaving some companies extremely exposed to price competition. Machines with similar functionality can run from $1,000 per line to $20,000 per line. The cost of memory or storage capacity in particular is declining rapidly, and will accelerate with the advent of read-write optical disk systems. Joint marketing ventures will proliferate as vendors recognise the benefits of selling into new applications areas.

Service bureaux will blossom over the next two years, both for voice messaging and interactive voice response applications. Service bureaux permit companies to try out a system before adopting a system or an application to make sure it is really functional for the particular situation. In addition, by renting capacity on a larger machine, companies can experiment in one-time *ad hoc* promotional schemes for information dissemination with no up-front equipment costs. LINK/NTV also expects off-line audiotex service providers such as Dowphone to sell capacity on their large scale machines to companies with similar applications.

Enhanced 800 services. Companies will begin adding interactive telephony technologies to their inbound and outbound WATS services to make them more efficient and capable of offering services not available otherwise. The phenomenon of being put on hold may soon disappear completely with sophisticated call handling; and order-entry functions will speed the process of calling an 800 number immensely.

New technologies will enhance all the current equipment. To mention only two, artificial intelligence will greatly increase the sophistication of the systems' interactive capabilities, and compact disc (CD) technologies will dramatically increase the storage capacity of available equipment.

13
Talking yellow pages

Leon A Ferber

Figure 13.1

What is it?

Talking Yellow Pages is a misnomer. Technology has created a new medium that combines and intensifies the strengths of traditional broadcast, print and direct mail media. This new medium combines the reach and selectivity of a telephone directory, the timeliness of newspaper advertising, and the power and quality of radio advertising to provide consumers with an audible network of informational services accessible through the telephone. Talking Yellow Pages is rooted in the ability of telephone directories to reach the masses and serve as locators and buying guides and carry advertising aimed at exact territories.

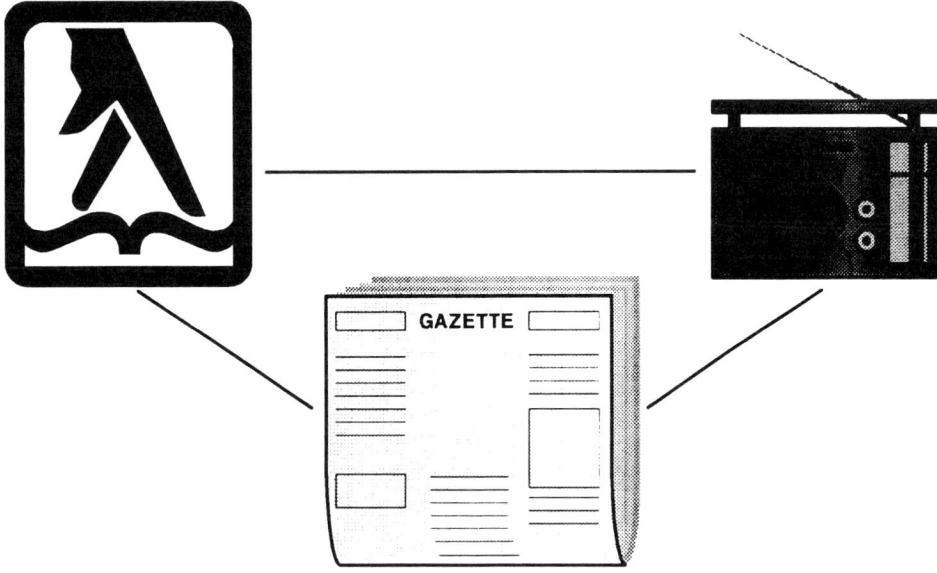

Figure 13.2

According to the 1985 National Yellow Pages Usage Study, 76.5 per cent of all adults referred to the Yellow Pages during the typical month. In the first quarter 1987, 82.5 per cent of all references resulted in a contact with a store or business.

Audio Response builds on this solid foundation and lets Talking Yellow Pages function more like a newspaper, responsive and timely, making it possible for advertisers to promote current specials and pinpoint their messages to highly select markets in any of a variety of categories or geographic locations, and more like radio, to let Talking Yellow Pages deliver high-quality promotional messages without the waste of radio spot sales.

Talking Yellow Pages is an inexpensive and infinitely flexible medium, adaptable to every type of message, to any time of day or night, and to every class of listener. Because the advertiser targets the customer directly and is in constant control of the message, Talking Yellow Pages is the most effective advertising medium available today. And it is local advertising—delivering the message in the local arena where most sales are made or lost.

Talking Yellow Pages provides a clear information channel for an advertiser's message. Capitalising on the intimacy of the medium, Talking Yellow Pages replaces the visual element of print advertising with sound effects appealing to a caller's imagination. It removes the clutter and the need to compete with other advertisers for attention in a newspaper and avoids the vagaries of listeners tuning in on a crowded radio dial. Importantly, Talking Yellow Pages borrows some technological resources from television, too, offering the equivalent of television's instant replay, with its ability to let a caller manipulate time by running a message again, skipping over a repeated one, and stopping it.

Accessed through the ubiquitous touch-tone telephone, Talking Yellow Pages is the single source to provide immediate and specialised product and service information to many callers simultaneously. In addition the service has the following advantages: control of the service is centralised; production is local for advertisers and automatic for sponsor-serviced messages and public service announcements; and usage is on demand over the telephone.

Who needs it?

For the local advertiser, Talking Yellow Pages represents a unique way to increase traffic, turn over inventory, and bring in new customers right away. It adds reach to any advertising campaign without the waste. Local advertisers will use Talking Yellow Pages in many ways. Some may choose to reinforce regular price-line advertising. Some may want to take advantage of changing seasons, fads and market demands and adapt their selling message accordingly. Some will place emphasis on special reduced prices, advertising items on sale to stimulate the movement of particular merchandise, while others may do clearance advertising to clear themselves of slow-moving product lines and make room for new lines or models. Whatever the message, these high-impact promotional ads are created to get immediate action. Their purpose is to stimulate sales. Maybe they will attract new customers into the bargain.

In short, any local advertiser looking to increase profits will be interested in Talking Yellow Pages. It lets a merchant sell his product or service with a unique audio-promotional message, providing the environment for a persuasive presentation of the product's special features without interference to the selling message.

At little cost, an advertiser using Talking Yellow Pages can expand his reach significantly and gather important demographic data to further refine selling strategies. By targeting specific audiences, advertisers can, of course, increase the efficiency of their commercials. Talking Yellow Pages provides true and exact customer and sales information including statistical usage analysis necessary for better marketing decisions. For the consumer, a walk through the Talking Yellow Pages becomes an interactive multi-media selling session—display ad, radio announcement, directory—and yields a wealth of buying information.

Wider consumer knowledge and a broader choice are the results for consumers who'll find their shopping time optimised by a telephone-based buying guide. For the price of a local call, it steers shoppers to businesses and provides them with current information on special promotions and sale items. Its ability to act as a locator service, providing the name and address of 'the store nearest you,' is another important feature.

Information providers are positioned to close the gap between sellers and buyers. An information provider that carries the advertiser's message is the vital connection between the merchant and the customer. This is the third link in the communication chain: information providers actually become the medium. Further more, opportunists will act as suppliers. An open-eyed information provider will be quick to become an independent production house specialising in audiotape production, and might include a research function as part of its efforts.

Advertising suppliers will be crucial to the success of Talking Yellow Pages. By assisting in the production process, information providers will create and fulfill a demand for scriptwriting, advertising advice, and the need for access to recording studios and voice talent for highest quality promos.

How does it work?

The successful Talking Yellow Pages service is one that conforms closely to the conventional ad development and placement process familiar to advertisers. Minimal disruption of the ad placement process will ensure rapid integration of this new service.

An advertiser wanting to utilise this medium simply contacts a Talking Yellow Pages representative (or vice versa for aggressive marketers). Working with the information provider, the advertising objective is defined, the selling message created and both print and audio production undertaken. The advertiser may submit production quality recordings if available, or utilise the production capabilities

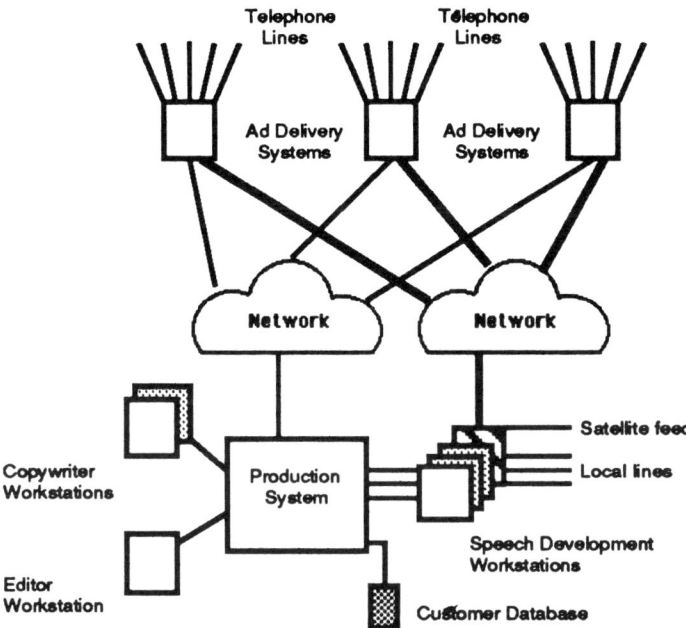

Figure 13.3 System configuration illustrates the four functional component systems interacting in the Talking Yellow Pages system configuration. They are Delivery, Distribution, Production and Accounting.

of the supplier. The last step is to assign the numeric access code to the account and feature it prominently in the advertiser's display ad. To alter the selling message is just as simple, requiring another set of objectives, a different selling message and another production session.

The consumer lets his fingers do the walking to the display ad and the dialing to the system, and his ears do the listening to the prerecorded promos. At any time the caller is able to repeat the message, stop the message or listen to another message.

Delivery

Delivery is accomplished through the use of a 'talking box' system architecture to provide the telephone line interfaces and the promo selection and playback capabilities, and handle the promo storage and network interface. The architecture is the result of careful human engineering which assumes the most important part of an application to be the user interface. Because busy lines must be avoided, the telephone interface is capable of supporting between 50 and 300 telephone lines at each remote site. Additionally, the system must support simultaneous access by many callers to different promos and allow the caller to select, play, repeat and skip over promos. To service this many callers and provide this amount of flexibility, storing 100 hours of promos on a delivery system is common.

The disc-based storage system accommodates digitally-encoded speech in order to provide the highest quality audio output, whether it is a simple spoken announcement or a professionally produced promo loaded with special effects. The promos will be loaded into the system in advance of the run date so the system must be able to select and run promos as scheduled based on a time stamp.

AD DELIVERY SYSTEM(s)

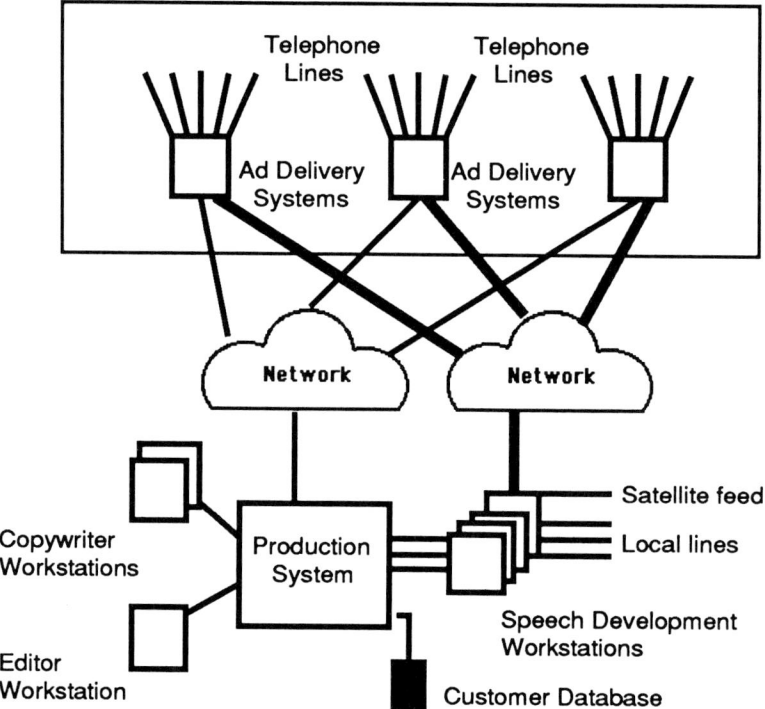

Figure 13.4

The network interface lets the independent groups—advertisers, consumers, information providers—interact. This interface is responsible for communicating command and control information from the Production system and for receiving time critical promos such as weather reports and financial news. The Delivery system—through the network—must be capable of simultaneous, two-way communication to the Production system while supporting telephone lines without degradation to caller services.

Production

Production methods for advertiser promos and sponsored services differ. For the advertiser, production begins with the script. An advertiser can use to his advantage a quick and simple 'live' promo which requires only sending the selling message and a recording of any music or special effects to Talking Yellow Pages to be copywritten and edited, and then recorded by an announcer on a speech development workstation. If preferred, the advertiser can submit a prepared promo in the form of prerecorded tape. New messages are as easy to create. The advertiser submits the new selling message and copywriters and editors prepare the text for recording by a staff announcer. Prior to distribution, each promo is reviewed by the editor and then written to a distribution floppy disc.

Sponsored services are produced directly from information sources such as news reporters, weather reporters, and business news wires. To create a sponsored services promo, an authorised source phones

AD PRODUCTION SYSTEM

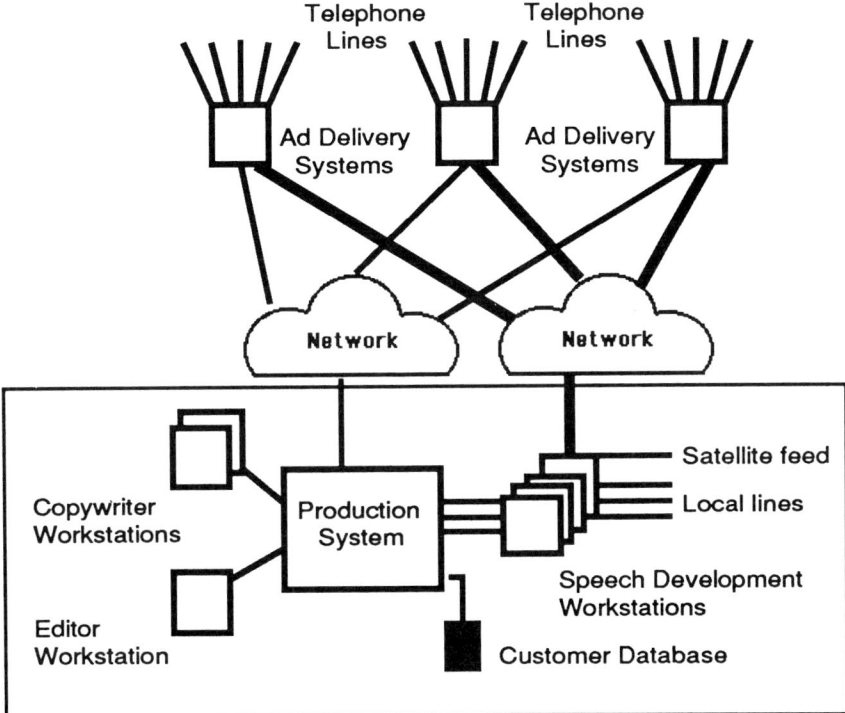

Figure 13.5

into a dedicated speech development workstation and enters an authorisation code. Validation of the code lets the source make a report over the phone. This information is digitally encoded as it is received and stored on disk. The message is then spliced to include a sponsor acknowledgement, header or trailer or both, distribution is assigned and the message is transmitted over a voice grade telephone line to all remote Delivery systems.

Distribution

Distribution of the promos can be accomplished by delivering duplicates of the floppy disc to remote sites, or expedited through the network, depending on how time-sensitive the information is. As with conventional advertising, most promo placements will be planned in advance to coincide with other marketing activities. With proper planning, promo distribution may only occur weekly and a physical medium, eg, disc or magnetic tape, will be used. These distribution media, specific to their area, are shipped to the remote delivery system sites for loading during off-peak hours.

Because of their critical nature, sponsored services will typically utilise electronic distribution. Updates are sent immediately over the network to all delivery systems which monitor the network and intercept only those messages with their location in the distribution header. When a remote site detects an update, the delivery system stores the message on disc by over-writing a previous update.

SPEECH DEVELOPMENT WORKSTATION(s)

Perception's Speech Development workstations create new messages and modify existing messages. The workstation is menu driven and provides for on-line audio digitization. The system accommodates both local announcers as well as remote dial-in for sponsored services such as news, weather reports, and public service announcements.

AD RECORDING:

Menu-driven
On-line Digitization
Create New and Modify Existing Recordings
   with Input from:
      Microphone (in-house announcer),
      Telephone (remote announcers),
      Tape (recording studios),
      Satellite (news wire services)

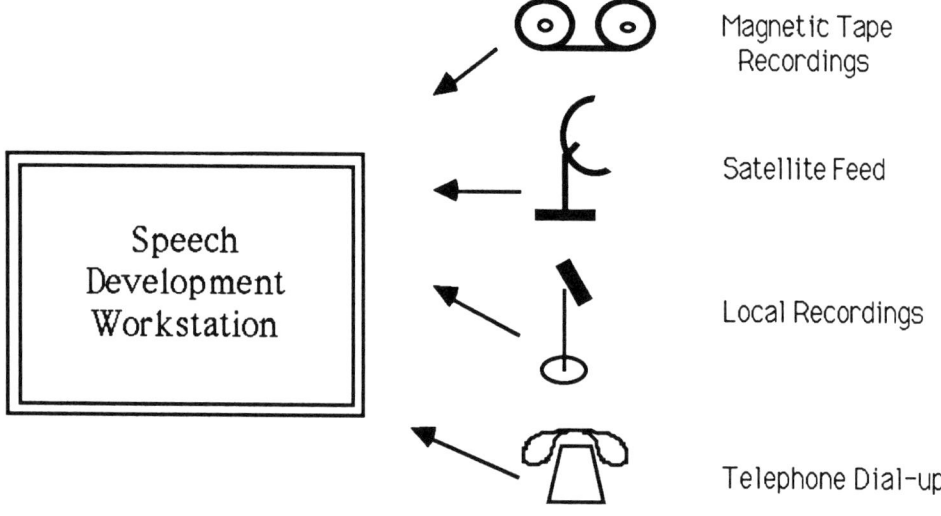

Figure 13.6

Accounting

The statistical usage log which tracks promo access and telephone line usage is not only important for accurate client billing and system planning, it is a distinguishing feature of the Talking Yellow Pages medium which can be used to attract advertisers.

The advertiser benefits from greater accountability than is offered by any other medium, pinpointing how many times a certain promo was accessed, at what times, and from which locations. Advertisers can test consumer response to a product promo during different seasons of the year, days of the week, or hours of the day. They can test whether frequent updates to the selling message are more effective than a one-time message, and can determine to what extent the message itself was effective.

Finally, for the advertiser who wants to measure overall results to evaluate the extent to which advertising accomplished its objectives, the usage statistics provide the most control of any medium for the advertiser. Currently, more than 700 lines of Talking Yellow Pages are running in the US. In Europe, several dozen lines are active. A typical system price starts at under $100,000. Each Talking Yellow Pages provider sets its display advertising rates independent of the audio services. It is product differentiation—not price—that attracts advertisers.

Summary

Talking Yellow Pages is a new medium revolutionising local advertising. Combining the local emphasis and reach of a telephone directory with radio's strength to change a message quickly, it retains a sense of immediacy and offers quick response for consumers and easy accountability for advertisers.

14
Data that talks and calls people up
Paul F Finnigan

We usually have to chase information. It does not chase us. The information required, too, can be difficult to find and sometimes impossible to find in time to use. For an investor, for instance, information that is a minute old may be too old; the same can be true in an airline emergency. Physicians, of course, know that seconds are precious. What if information, first of all, were readily available electronically—so readily available it could talk to us? Even better, what if the information we needed could actually call us as soon as it became available? It can. Voice mail/audiotex is the method.

The world is full of information. The world is also full of people who need information. The world is also full of computer databases with the information that people need. While conditions might seem perfect for matching needs with what is needed, they are not—not without voice mail/audiotex, or, in other words, not without databases that can talk and call you up. The world is not full of computer terminals to access the information in databases, but it is full of telephones. Telephones are all you need to access a computer database using voice mail/audiotex.

Information providers, computers, databases, and telephones are almost all we need to provide information services required by investors, travellers, buyers and sellers, and others who need information right away. However this is an over-simplification. These are the components. How are they successfully combined? Voice mail/audiotex can teach the database to talk, and even allow the database to call you up.

Investors

Considers investors, for instance. Investors want financial news as soon as it happens. They can get it by listening to the radio, by watching television, or by calling DowPhone—one of the early audiotex programs. For any of these previous methods of giving out information, investors have to be listening, or they have to call in. People do not always have time to listen, or they forget to call in, or when they do there may be no new information awaiting them.

The new generation of audiotex products, such as the predecessors of DowPhone, can automatically deliver voice mail or audiotex messages in natural voice. In addition they have the ability to call someone by telephone or, for time-critical information, by pager ('beeper') whenever something special happens.

To give an indication of the advantages, consider a financial adviser who wants to notify some of his investor clients regarding a change in his investment position on a certain company. The problem is that it takes hours of calling, telling, and playing telephone tag, and there is always the question of "who do I call first?" Audiotex changes all that. The information provider calls the voice mail computer and records a recommendation. The message is recorded under a particular voice mail box number. The box number will provide security, as well as tell the recipient certain things about the message. The information provider selects a group number (a list of subscribers who have signed up for news on this particular subject). The group number and box number are passed to a network paging service which identifies the pager number and location of each subscriber in the group and which in turn notifies the paging service in each appropriate city. The local paging service transmits the alert and the box number to the pager. The pager beeps and the digital display shows the box number. In addition to providing a key to the information, the box number may also denote urgency, industry, and even company.

Two hundred, or even 2,000 subscribers are simultaneously alerted to the investment news broadcast, which is then received by calling the voice mail service, entering a subscriber password and the box number displayed on the pager. The elapsed time for the broadcast is the time for the information provider to dial an 800 number, enter a password, select the box number and group number, record and edit the message, and initiate the page alert. After two minutes, five minutes or maybe ten, depending on the length of the message, beepers are going off in 35 cities!

Security is high because the message has high value. The receiver must have a seven-digit password (this can also be used to verify for delivery), plus a seven-digit box number, to retrieve the information.

Voice messaging

While delivering information in this manner may sound great, the fact is that recording large amounts of information in this manner can be impractical for the kind of information that the investor, the airline, the traveller, or the buyer or the seller really wants. The key to delivering lots of information over the telephone is that the data must talk. Thanks to voice mail/audiotex, any information residing in a database can now be available over the telephone, at any time from any part of the world.

Voice mail converts analogue voice to stored digitised voice, which it can then manipulate in audiotex applications. A telephone line comes into a voice mail computer through telephone line coupler cards—a link with the outside world, regulated by the Federal Communications Commission. The coupler cards connect with line cards, which recognise the touch tones from the caller's telephone, and digitise the incoming analogue voice and sounds. The line cards connect wih the computer cards, the equivalent of a minicomputer, that contain all of the logic for assigning addresses, storing messages, forwarding messages, and issuing or recognising commands. Disc storage devices actually store the messages and the programs for the entire configuration. The operator station cards allow the voice mail computer to intercept to a live operator when touch-tone is not available or operators need to provide direct assistance to callers.

In the example above of the pager and the single message, it was shown how the information is recorded into a stored location. The voice mail computer, working with a company's database, allows an additional step where the caller can directly access a variety of stored, digital information.

Audiotex designers can store a virtually unlimited vocabulary in the voice mail computer—words, sentences, paragraphs, stories. Each one of those has a box number (a location number). Through the audiotex interface, the voice mail computer matches incoming ASCII data from the company database with stored voice: words, paragraphs, and sentences. It delivers those stored voice words as total, natural voice messages.

For the investor, the database might indicate 'ACME 1349763.' The audiotex computer translates the data to messages 1, 3, 49, and 763. The voice mail computer matches the incoming messages with stored voice messages, and delivers the information to the caller "ACME Up $\frac{1}{8}$ to $8\frac{3}{8}$."

Voice messaging applications

For investors, then, audiotex solves the delivery problem between a database and financial people. Airlines are finding applications, too. Take, for example, the audiotex product outlined below where a company needs to notify several thousand employees, in this case flight attendants, about their flight assignments.

TWA faces a problem common to all airlines, daily scheduling of reserve flight attendants—crew members called, on short notice, to meet the dynamic situations caused by weather, holiday travellers and schedule changes.

TWA has a harried switchboard as 'reserves' call in for their assignments. The TWA reserves start by making a local call and listening to an answering machine deliver a list of the names of those with assignments. Many get engaged signals, because there can be up to 100 people trying to call through at the same time. Those included on the list then call an 800 number for their specific assignments, often having to wait on hold for their call to be handled.

Enter audiotex! TWA has decided to implement an audiotex application to provide the solution.

The crew member dials an 800 number for voice mail.

Voice mail answers and says, "Hello, this is TWA Crew Schedule. Identification number please." After receiving the identification number, voice mail asks for an additional password.

Once it receives the password, the voice mail computer communicates with TWA's crew management computer and enters an inquiry with the identification number and password. The TWA computer responds, in digital form, giving the assignment. Based on the information from the TWA computer, the voice mail computer concatenates a complete voice message from the stored vocabulary.

Voice mail delivers the message: "You have been assigned . . . pairing . . . four . . . two . . . three . . . one . . . departing . . . Flight . . . one hundred . . . fifty . . . August. twenty . . . eleven hundred . . . hours local time . . . your lay-overs are . . . LAX . . . MCI . . . and . . . STL . . . . To OK your assignment, press one, Otherwise, press nine."

If the caller presses one, the voice mail computer delivers a notice to the crew scheduling computer that the person has received the message and accepted the trip. If the caller presses nine, he or she is switched directly to a 'live' scheduler. If there is no assignment, the computer simply says, "You are released until Monday at 7 pm" (the next call-in period).

The TWA scheduling department can enter assignments at a CRT, and voice mail/audiotex delivers them over the telephone. The employee enters an identification number and a password. The computer inquires and retrieves the data and delivers the information. Callers listen to information specifically for them, and to nothing more. They make one call instead of two. Voice mail audiotex cuts down on busy signals. TWA expects the average connect time to be half that of a manually handled call.

In another airline application of audiotex, Pacific Southwest Airlines offers flight schedules, and arrival and departure information by telephone. Passengers use a touchtone phone to answer computer prompts for their 'from city' and 'to city.' The PSA computer searches the database for the routes between those cities matches information with voice messages, and delivers flights and times over the telephone.

Voice mail audiotex can deliver medical updates in much the same way it delivers flight information. Every US physician now has access to audiotex that supplies updates to the *Physician's*

Desk Reference (PDR). The PDR is published annually, but the physician may need more recent information than the printed source provides. To receive current updates about drugs (interactions, indications, and other specialised medical information), doctors call in from their own telephones and key in a personal identification number when asked for it. Then, referring to a listing of code numbers for various drugs, the code number for the drug being researched is keyed in. Voice mail audiotex then delivers an update by telephone.

Telemarketing

Voice mail/audiotex also advertises goods and services by digitised voice and can take orders directly from the telephone into the computer. In this case audiotex amounts to a voice bulletin board. A person calls in to the voice mail/audiotex computer and listens to a talking bulletin board of discount items being offered for sale. If the caller hears an item that he or she wants to order, he presses the star key, which puts him into the voice mailbox of the telemarketing service.

The caller can then place an order with the service, speaking a message directly into the mailbox of the service, a voice mail-order service. Or the caller can place a direct phone call, outside audiotex. The service tells the caller where to get the item, or takes his order over the telephone and arranges delivery. One Los Angeles entrepreneur, for instance, had 200 hours of caller time to his bulletin board in one recent month: at an average of one minute per call, that means 14,000 transactions. As a voice mail-order house, he does not even have to own a store. He takes orders, and sends out purchased items by United Parcel Service.

Voice mail is one component in what the end user experiences as audiotex. The new ingredient is the audiotex interface between the host computer database and voice mail. In developing the audiotex applications themselves for TWA and others, the voice processing industry has drawn on previous experience and accumulated new experience on what works and what does not.

Informal tests of the people using audiotex have found, for example, that the most effective way to give instruction in an audiotex application is first to tell people they are going to have to make a choice, then list the choices, explain each choice one at a time, and finally ask for a decision. Three is a good number of choices to give, and six is a maximum before people just cannot remember the options well enough to make good decisions. People tend to remember the first and last choice but not the ones in the middle.

In an audiotex Adventure game, for instance, the narrator's voice would say, "You can be a hobbit, or an elf, or a dwarf." The voice would list the advantages and disadvantages of each, then say, "Now make your choice. To be a hobbit, press one. To be an elf, press two. To be a dwarf, press three".

The voice processing industry, then, has learned from experience what works and what does not. Audiotex does not simply deliver. It delivers with logic and feeling. And the wording of the choices can include some hefty, spoken selling—giving audiotex a feature not available with videotex for motivating people to action. Taking the Adventure game example, you say "You can decide to be a hobbit if you want to, but you'll have a short life span that way. Most people have better luck being a dwarf."

Perhaps the biggest difference between audiotex and videotex, though, is simple. There are telephones in every airport, restaurant, apartment, and living room. When did you last see a public computer terminal? You may some day, but not very often today.

PABX and Voicemail

No one using audiotex, of course, has any more idea that they are using voice mail than that they are using an IBM mainframe or coaxial cabling or the public switched network. Audiotex is a product. It is simply a service, as transparent to the user as the poles and wires connected to the telephone.

During the early 1980s the voice mail industry was not selling such products but, rather, was trying to sell voice technology itself. And the industry was engaged in a features war. Competing vendors dressed up basic store and forward service with literally hundreds of features. Some options were valuable, but observers wondered if it was really necessary to have pause, skip forward, skip backward, and other capabilities that tended to make voice mail complicated to use. Who was going to benefit from this service, vendors needed to ask, and what features did users realistically require?

The industry spent millions of dollars on promotions, and the message was always the same, "Eliminate the pink slip". Everyone saw voice mail as a form of office automation. Most of the voice processing industry was working on Public Automated Branch Exchange (PABX) integration. Many articles sighted the near-term market for voice mail as the PBX. The PABX is basically an automated switchboard. Linking voice mail to the PABX provoides an automated answering machine within the office. To some industry insiders, though, the question remained, "So what?" An automated answering machine is a long way from services like crew scheduling, or, now, instant investor notification.

Connecting with someone's PABX is an opportunity, but, some businesses are beginning to think, so is connecting directly to the company's computer so that any computer can be made to talk. And now the industry is building larger configurations of computers that connect a voice mail/audiotex computer directly to the public switched network. The central office then become's voice mail's PBX.

New voice mail technology will handle thousands of calls simultaneously today. The voice mail computer has 128 lines. By configuring additional, identical computers together, the linked voice mail computers can handle many more lines. For example, ten linked voice mail computers can provide a 1280 line switch.

One-way communications

Having voice mail available to the public, of course, is not the same as having the public use it. The only problem for audiotex, in fact, has nothing to do with the kinds of technical problems that voice vendors have had more than enough time to resolve. The remaining hurdle is getting people to use the technology.

The majority of the traffic on present voice mail services, for example, is one-way communications, not two-way. People are willing to listen, which is why DowPhone, TWA crew scheduling or PDR works so nicely. To use it, all people have to do is call in and listen. Experience is beginning to break down the old barriers, though. As people become experienced with voice mail, they become senders.

The voice mail industry, offering what a lot of people saw as an overgrown answering machine, has sold only a few hundred complete voice mail configurations since its beginning in 1980. Many companies have concluded that messaging is not enough, even at as low a rate as $20 or $30 per person, per month. It may be that the people who really need voice mail are not in the office, and that those who are in the office really do not want to replace their pink slips. It may be that we need voice mail with audiotex, to improve the economics so that users can justify the cost. When voice mail becomes audiotex, people use it to send recorded messages, listen to news reports, plan investments, and, in short, to do business.

Voice mail/audiotex, then, could become as common as the telephone or, at least, as the company database. Almost every company has a computer and a database. Think what audiotex can do for the travel industry: 70 to 80 per cent of all business travellers change their itinerary at least once on every trip. When they change, someone with a telephone and a CRT has to access the huge database that the travel industry maintains. Audiotex could eliminate the middleman. When travellers pick up their tickets, they could be given a boarding envelope with an audiotex 800 number. To change their flight, they listen to the flights and times available, and make a choice—date, time, flight number.

The financial industry is another good example. There is so much constantly changing information on stocks, bonds, prices for gold and silver. Today people sit at CRTs with a telephone dispensing the information. Audiotex is changing that.

Even the giant, timeshare databases are finding voice mail/audiotex. Dow Jones already has it. The Source or CompuServe and others will eventually be available by telephone. People think they need a computer terminal to access a timeshare database. All they need is a telephone.

Data that talks and calls people up

Nor, as mentioned earlier, does audiotex have to end with the simple dispensing of information. It can call out. In fact, with the computer's "if this, then that" capability, business people can have extensive computer capabilities available to them over the telephone without a middleman.

Right now, for example, a voice mail subscriber can use the telephone keypad to send a deferred message. The computer asks the sender to answer questions. The subscriber goes through a series of simple steps to indicate delivery date, time, number of tries, and re-try interval. "Tomorrow, eight thirty, am . . . four tries every fifteen minutes."

In the next generation of voice mail software, users will be able to set up comparable if-then conditions for the computer. "If my voice mail stock is $200 a share," the investor can instruct the audiotex service, "then page me as soon as it happens."

In a more advanced application, the businessman could specify, "If the composite of my stocks goes up or down more than three points, call me." Again, the application would reside in the computer. "Touch five to set up notification on stock prices," the application would begin, and then a series of questions would lead the caller through the steps to program the computer.

The final question from the computer would be, "Enter telephone number where you are to be notified, including area code." The investor doesn't even have to call in to find out what is going on. He will know if his portfolio moves significantly in either direction, because audiotex will call him— every hour on the hour at up to six different locations if necessary. Or, it will page alert anywhere in the country or the world.

What are other possibilities for the forthcoming if-then capability? Voice mail will automatically check travel services for specially-priced vacation trips; check executive toy lists for Porsche speedsters, or airplanes or anything, and call you when it finds one.

When people used to hear about voice mail, they would often ask, "Why do I need it, why not just call someone if I want them?" For each objection there was an answer. "If you know where they are, you probably should call them. But if they are out or in a meeting or travelling, then maybe you should try voice mail." Now voice mail/audiotex provides a new answer. "Because now you do not have to worry about whether you should call or not; when the time is right, voice mail/audiotex will call you."

15

Optical storage media: CD-ROM and beyond

David Shorrock

The development of CD-ROM

Optical discs first appeared on the consumer market in the early 1970s. Companies such as Philips of the Netherlands and MCA of the US, developed systems for bringing pre-recorded television programmes into the home. The systems were similar (but incompatible) in that they both used a 12 inch diameter silvered disc, that was spun at high speed. The recorded information, held as small spots or pits in a spiral track, in much the same way as a conventional LP record is read by a tightly focussed laser beam. This information is then decoded from the variations in the reflection of the laser from the surface of the spinning disc.

Unfortunately, the laser video disc had missed the window of opportunity in the consumer market: the video tape recorder had arrived, and consumers saw the benefits of a system that could record and 'timeshift' television programmes as well as being able to play pre-recorded tapes. After a few short years the videodisc disappeared from the consumer market, although a small number of systems are used in commercial or educational applications, (for example: the British Broadcasting Corporation's Domesday Project).

The failure of the videodisc, particularly the lack of standards and the plethora of incompatible systems, was heeded by the manufacturers, and in particular, Philips and Sony, who collaborated on a new product, the compact audio disc or CD, (or CD-A, compact disc audio, to distinguish between other types of CD). The compact disc was smaller than its predecessor, 12 centimeters (4.72 inches), made of metallised polycarbonate 1.2 millimetres thick, and with a spiral track some three miles long wound to within 1.6 thousandth of a millimetre. Unlike conventional records, to maximise the playing time the speed of a compact disc varies from 200 to 430 rpm, although the speed at which the digital audio information passes the laser 'stylus', the linear velocity, is constant, (referred to as Constant Linear Velocity).

In stark contrast to the failure of the videodisc, the compact disc has been a commercial success. When first launched in 1983, sales worldwide of compact disc players were approximately 350,000, with 1.25 million discs. By 1986, this had grown to 14 million players and 100 million discs, with predictions of 300 million disc sales in the US alone by 1990.

From recording digital audio on a compact disc, it was a short step to recording computer data. Philips and Sony continued their collaboration with the development of the compact disc read-only memory (CD-ROM). A single CD-ROM is capable of storing up to 600 megabytes of data, the equivalent of a quarter of a million printed pages of text, the entire contents of the *Financial Times* for three years.

However, it is the vast storage capacity of the CD-ROM and the comparatively low cost of volume reproduction that is creating the dilemmas. What do I store on it? How do I retrieve the information? What is the market? And perhaps most importantly, how can I ensure that the disc I make can be read by others? The answer, at least in part, lies in establishing standards.

CD-ROM standards

The Philips-Sony collaboration has benefited the whole compact disc industry. To avoid the debacle over competing and incompatible standards that surrounded the videodisc, Philips and Sony have made their standards public and offered licences for other manufacturers to follow. These standards are contained in a series of 'coloured' books: the *Red Book*, which defines the physical standard for (primarily audio) discs and players; the *Yellow Book*, which extends the standard for data on CD-ROM.

The physical standards are the basic standards for CDs. In the case of CD-ROM however, the standards must extend further into the 'logical' level, the definition of common file structures of records, files and volumes and their respective directory in the database. Without such logical standards it will not be possible for CD-ROMs to be used on differing manufacturers' computer systems.

One of the most prominent forces for logical standards has been the High Sierra Group (HSG), named after the hotel in Nevada in which the inaugural meeting was held in 1985. The group consists of both manufacturers and software companies with interests in CD-ROM development, but interestingly, no information providers. HSG includes Apple, Hitachi, Microsoft, Microware, 3M, Philips and Sony. Noticeable by their absence (they declined an invitation to participate) is IBM, who, some observers believe, could well introduce their own proprietary standard at some future date.

Subsequently, HSG received contributions from such bodies as the European Optical Disc Forum, the US Information Industry Association and the American Library Association. In 1986, HSG published a draft standard, which is being sponsored by the US National Standards Organisation and is on its way to becoming accepted internationally and approved by the International Standards Organisation. The HSG standard defines a common volume, file and directory structure, while avoiding constraints that might inhibit future developments of CD-ROM. Working to this standard, information providers can publish a CD-ROM which can work on any drive, on any computer, or microcomputer system.

One of the potential inhibitors of a mass market for CD-ROM products is the search and retrieval software that extracts the required information from the disc. Fast, free-text retrieval software has been developed for the on-line database industry and has been available for a number of years. With the advent of the microcomputer, vendors have transferred their products and now see the CD-ROM as a new opportunity. The choice of software, because it is prorietary and, as yet, there are no standards, has major implications for the CD-ROM publisher because it influences the design and indexing of the database. A database structured for use with Battelle's MicroBASIS for example, cannot be interrogated by BRS/Search. The implication is that the choice of search software is one of the key decisions to be made by the publisher. The other, of course, is what database to put on the CD-ROM.

Applications of CD-ROM

The first prototype CD-ROM was demonstrated in 1984. By the end of 1987, it was estimated that there were some 400 CD-ROM products available, although the majority of these are small pilot systems. The following are examples of some of these systems.

Grolier: the Academic American

The US publishers Grolier claim to be the world's biggest publisher of encyclopaedias. Their main product is the 20 volume *Academic American Encyclopaedia* which is found in thousands of American households. In the early 1980s, Grolier set up a subsidiary, Grolier Electronic Publishing, to make the *Academic American Encyclopaedia* available as an on-line database. In 1986, after an abortive attempt at offering a videodisc-based encyclopaedia, Grolier launched the Academic American on CD-ROM. Since it's launch, the first issue has sold over 2,500 copies at $199 and a second edition is now available for $299. The Grolier disc supports a range of CD-ROM drives, but under the IBM PC (or equivalent) operating system MS-DOS. In order to capture the American school market, which has traditionally favoured Apple computers, Grolier have recently made available an Apple version of the disc.

The Grolier search-and-retrieval software is the Knowledge Retrieval System (KRS) by KnowledgeSet, which has powerful search facilities as well as being easy to use and visually attractive. One of the special features is a browse facility that allows the user to browse through the database in much the same way as one might through a book, stopping and exploring in more depth at an item of interest.

Microsoft's Bookshelf

The PC software company Microsoft have launched a $295 'Bookshelf', a CD-ROM reference library which includes Roget's *Thesaurus*, the *American Heritage Dictionary*, Bartlett's *Quotations*, a word almanac, spelling checker and style manual, as well as a directory of postcodes. The power of Bookshelf is that it can be integrated into all the commonly-available word processing packages available on the IBM PC. Now, instead of having to pause to look for a reference, or check the meaning of a word, the user can call up the relevant reference with a few keystrokes and, if he wants, copy it into his own document.

A CD-ROM atlas

DeLorme Mapping Systems of Freeport, US, have demonstrated a world atlas that allows the user to select a particular map and then zoom in. The demonstration disc is a pilot to show the power of the media and to tempt potential customers. The zoom feature allows the user to locate a given country and then continue zooming through districts, to towns and then down through to streets and even individual houses. The demonstration also includes three-dimensional images of elevations and ocean features.

Whether the DeLorme atlas becomes a commercial reality remains to be seen. It may never become a consumer product, but it has certainly attracted the interest of the military, who have long appreciated the importance of mapping and the ability to use the terrain for tactical advantage.

ADONIS

ADONIS is a trial document delivery service that supplies some 219 biomedical journals on CD-ROM. The discs are delivered at approximately weekly intervals to major document supply centres in Europe, the US, Mexico, Australia and Japan. The project is financially supported by the Commission of the European Communities and the ten cooperating publishers. The aim of the ADONIS project is to reduce the cost of copying and distribution of information for lending libraries.

Each week the contents of the journals are indexed by Excerpta Medica in Amsterdam; this index information is then merged with a facsimile of the journal articles by Scanmedia Limited in the UK; the facsimile and indexing information is then mastered and the discs produced at the Philips and DuPont Optical plant in Hannover; from here, the discs are dispatched to the document supply centres.

The information from the discs is recovered by means of the ADONIS workstation, comprising an IBM PC, a high resolution monitor, proprietary cards that handle the facsimile image compression, a CD-ROM drive and a laser printer.

The major limitation of the ADONIS project is that the text of the articles is in facsimile rather than textual, 'machine readable' form. The latter would allow more intelligent access to the disc, for instance, by word searching. This limitation reflects the use to which the media is being put: a substitute, on-demand printing press, rather than the mechanism for information delivery with which the end-user can interact.

The future: CD-I, CD-V, DV-I or WORM?

CD-ROM as a media is still at a formative stage but, just as its audio counterpart is being threatened by digital audio tape, (DAT), CD-ROM faces the challenge of an intelligent offspring, compact disc interactive, CD-I.

Philips and Sony demonstrated CD-I at an industry conference in 1986. Its advantages over CD-ROM are described in the Philips publication *CD-I: the Compact Disc goes Interactive* as "its ability to supplement the basic text and data information with visual material like still pictures, diagrams, high-quality computer graphics and cartoon-style animations, plus sound of every conceivable kind from top quality stereo to speech and sound effects, all simultaneously".

So we see that CD-I is a multi-media product. Philips attempt to differentiate between CD-ROM and CD-I by saying that CD-ROM is a software and data standard, while CD-I is a system standard which defines the player and the player software, as contained in the Philips/Sony *Green Book*. This distinction is somewhat artificial and is perhaps at best an attempt to position the two at different markets: CD-ROM as a computer peripheral for professional users, CD-I as a stand-alone product with potential for the leisure market.

The CD-I player is a complex device that includes a powerful Motorola 68000 based, 32 bit microprocessor and over a megabyte of memory. Behind the hardware is a sophisticated software Compact Disc Real Time Operating System (CD-RTOS) which handles the switching of video and audio from the disc. But despite the microprocessor-based hardware, the CD-I player will resemble an audio CD player, with the controls confined to a simple handheld infrared unit.

Philips and Sony intend to arrange a series of demonstrations of CD-I during 1988 with a commercial launch scheduled for early 1989, after which sales of 30 million units a year are predicted. Time will tell as to the accuracy of this market projection but, from the demonstrations of pilot systems, CD-I certainly looks an attractive and entertaining media.

Unfortunately, Philips/Sony appear to have developed a competitor to CD-I before it is even in commercial production: Compact Disc Video, CD-V. First shown in 1987, CD-V is an extension of the audio compact disc, combining 40 minutes of digital sound with four to five minutes of full-motion analogue video. CD-V has yet to make an impact as only demonstration players and discs have been shown. Given the limitation on the amount of video material that can be stored on the disc, CD-V is either just a clever demonstration of the technology, or the pop video of the future.

A potential competitor to the Philips/Sony CD-Is and CD-Vs is DV-I, Digital Video Interactive, courtesy of RCA and General Electric of the US. Announced in March 1987, DV-I offers an hour of full-motion digital video and audio on a standard CD-ROM disc. The heart of DV-I is a special chip which contains the complex data-compression algorithms for packing the video, text and audio information onto the disc. RCA/GE expect to have initial production versions of the chip available in early 1988 for incorporation into an IBM PC. The initial price is expected to be several thousand dollars, falling to a few hundred within three years. Whether or not DVI competes with CD-I and

CD-V will depend on the approach RCA/GE adopt to licensing other manufactures. At present, DVI is a proprietary development, whilst CD-I and CD-V are agreed and published standards.

The development of new CD formats continues apace. In November 1987, Philips/Sony announced standards for CD-WORM, a 'write-once, read-many' disc. CD-WORM allows users to write their own data to a disc, which can then be read back any number of times without loss or deterioration. Once the data is written, it remains for the life of the disc and cannot be erased. The Philips/Sony CD-WORM is capable of storing 600 Mbytes of data and is being promoted as a personal computer peripheral, suitable for professional applications requiring non-erasable media, such as financial audit trails. The Philips/Sony CD-WORM is expected to be commercially available between late 1988 and early 1989 although, in the US, products are already available (in limited quantities) from IBM, Information Storage, Maxtor, Optimem, Optotech and Pioneer.

On the horizon is the erasable CD (CD-E?), which can be written and erased in much the same way as the magnetic floppy discs with which we are now familiar. Companies such as Energy Conversion Devices, Matsushita, 3M, Sharp, Verbatim, Quantex, and of course Sony and Philips, are all working on devices which are expected to be commercially available by 1990, although evaluation units should be available in late 1988.

	Year	Units (000s)	Sales ($M)
CD-ROM	1988	250	150
	1990	690	150
CD-WORM	1988	50	100
	1990	300	210
CD-Erasable	1988	1	5
	1990	70	140

CD Drive Market Forecasts

Figure 15.1

The optical storage market

The developments in optical storage products based around the Philips/Sony CD format will continue. Sets of common standards ensure the development of a coordinated market. To the end user they are a guarantee that the product will be supported and the ability to exchange media with other users, or in the case of pre-recorded media, to have access to a wide range of material. To the information publisher supplying that pre-recorded material, standards offer a large and contiguous market. To the manufacturer, they offer a single homogeneous market, but they also pose a threat.

Manufacturers who go it alone (who, for example, have entered the market early and have products in advance of the standards) will see their market share fall. Philips themselves illustrate this problem with their Laservision 12 inch videodisc and their Megadoc system based around 12 inch WORM discs. Philips' own CD developments threaten the viability of these products.

The market for CD-ROM and its derivatives will continue to develop (and Figure 15.1 shows one forecast, courtesy of Communication Publishing Inc), but it will only do so if it is based on globally accepted standards.

16
Database distribution on CD-ROM
Graham Seddon

The on-line industry

BRS is one of the world's leading vendors of on-line information. Over 30,000 subscribers have instant access to the information needed to support research and decision-making activites in virtually every major discipline—health, medicine, pharmacology, the biosciences, science and technology, education, business and finance, the social sciences, and humanities. Via a terminal, a personal computer, or a word processing system, a modem and a telephone line over 50,000,000 documents stored on the BRS central computer can be searched to find the precise information of interest. The so-called BRS/SEARCH service offers a command language dialogue for the experienced user, and a menu-driven dialogue for the occasional user. In addition, full-text facilities combine the power of on-line searching with the convenience of document delivery.

The retrieval software used by BRS is known as BRS/SEARCH and has been sold to hundreds of customers throughout the world. In particular, it has formed the cornerstone of a number of on-line services in Europe—Datasolve, G.CAM, IPSOA, Bertelsmann, and Datacentralen, are just a few examples. Thus with some justification BRS can claim that its understanding of the on-line business is no more limited than anybody else's.

The BRS/SEARCH Service provides access to an immense electronic library maintained at peak currency; and it is the flexibility and speed at which this library can be searched which is the major attraction of on-line services. Entry to the library is free, but the user pays an exit fee based on time spent in the library (the connect-time), and the actual documents displayed (the document charge). Various subscription plans are available which offer connect-time discounts to those customers who make an advance commitment to using the service.

The type of information which can be offered by an on-line service in an acceptable format is constrained by current technology, in particular the bandwidth of the telecommunications link. Thus it is not generally possible to offer graphical or pictorial information. Initially on-line services concentrated on the delivery of bibliographic information which can be delivered quite satisfactorily over a telecommunications line. More recently, vendors have begun to offer full-text information (the source document as opposed to its bibliographic description) but the technological constraints still apply. It is not yet feasible to offer document images, and full text applications must be selected with care and require good editorial skills.

The characteristics of the on-line industry are summarised in Figure 16.1 in the context of conventional comparisons with the use of CD-ROM for database distribution.

Many would argue that the pay-as-you-go charging approach is an advantage to the customer, who only pays for what is actually useful. This approach is instinctively disliked by publishers however, who are used to selling complete publications, of which only a fraction of the contents might be

Advantages	**Disadvantages**
✳ **Currency**	✳ **Low Bandwidth**
✳ **Broad Subject Range**	✳ **Pay-as-you-go**

Figure 16.1 Characteristics of the on-line industry

relevant to a particular customer. People will often buy a book just to reference a single chapter. Pay-as-you-go is not an intrinsic feature of the on-line industry—it is perfectly possible to offer customers a single subscription charge which will give unlimited access to an on-line database. However, the pay-as-you-go approach has become traditionally associated with on-line services and is consistent with the concept of an electronic library. But the effect has been that on-line services are perceived as expensive and this has perhaps constrained their growth.

In all market areas, the ambition of the on-line vendor is to extend use of the system from the trained intermediary to the end-user. Much development effort is directed at making the service attractive to the end-user, via easy-to-use dialogues and full-text databases. However, the crucial barrier to further penetration of on-line services remains the low bandwidth of the telecommunications link which severely limits the type of information which can be distributed via on-line services.

Optical disc technology

BRS has long recognised the need to enrich on-line information and was quick to appreciate the potential of optical disc technology. In collaboration with leading-edge companies such as LaserData and Reference Technology, it has pioneered the exploitation of both the 12 inch videodisc and the 12 cm compact disc. For example:

> In 1982 the use of a video disc for containing pictorial information was demonstrated. The video disc was linked to a micro computer, itself communicating with a major on-line medical databases.
>
> In 1984 the use of a video disc for storing purely textual information was demonstrated. In this case the famous encyclopaedia.
>
> In 1985 BRS demonstrated CD-ROM technology at the Frankfurt Book Fair with one of the first discs to be filled fully with data.

Our experiences with CD-ROM technology is discussed at length below. But before that, it is useful to draw some general conclusions from our earlier experiments in order to set our approach to CD-ROM in the appropriate context. Two interconnected issues emerged of paramount importance in the exploitation of optical disc technology—Standards and Markets.

Standards

Even when used solely for storing pictorial information, our early experiences with video disc products were frustrating owing to the absence of international standards for TV signals. To display images from a disc manufactured in the US, the complete display system had also to be imported from the US—that is, it was impossible to locally acquire the appropriate monitors and players. When the video disc was used for storing digital information, the standards situation became even more frustrating. To compensate for imperfections on the disc surface, very complex error checking and correcting codes have to be recorded along with the data. Unfortunately these schemes were proprietary to particular suppliers and were anything but standard. Disc interchangeability was impossible. If you bought a disc from one supplier you needed to buy his hardware and software also. If you then wanted to buy a different disc from another supplier, you then needed to buy another set of hardware and software.

Markets

Closely connected with the standards issue is the question of the market size for any disc-based product. How many customers were prepared to purchase a microcomputer and videodisc drive just to be able to access the disc. Was the hardware easily available even if the customer would pay? And more significant, how many customers would duplicate their investment just to be able to access a disc from another supplier.

It also became clear that great care was needed to define what customers really wanted. Everybody agreed that pictorial information was an essential part of any information product, but surprisingly few people were prepared to pay for it when it came to the point. It also emerged that the initial market, certainly for the products of BRS, would be the reference librarian who would use optical disc technology as a cost effective substitute for the electronic on-line reference service traditionally supplied by BRS. Although released from the constraints of 1200 bps telecommunication facilities, there was surprisingly little enthusiasm for a less expert human interface. The reference librarian knew the BRS command language and was not passionately demanding a more user-friendly menu-structured dialogue with the computer. This observation should however be contrasted with the experience gained in other market sectors, where products directed at the end user had been introduced and had been very successful. Away from the domain of the reference librarian, a user friendly interface was absolutely essential and great care was needed to define that interface in relation to the experience of the user and the content of the database.

The above remarks summarise our early experience with optical disc technology, and in related areas where innovative information products had been introduced. It must also be stressed that the early efforts of BRS in optical disc technology were solely for research purposes and some important lessons were drawn. Technically optical disc technology was indeed revolutionary. But for its commercial exploitation a number of market dominated issues remained. The critical issue was entry-cost. Whilst the discs were cheap to produce, it seemed that the customer would not countenance purchasing a personal computer and disc drive just to access one disc, and the customer certainly refused to contemplate duplicating his investment just to access a disc from another supplier. Disc interchangeability was the single most important factor in market acceptance, and this could only be achieved by the adoption throughout the industry of internationally agreed standards. When the compact disc for audio applications was developed by Philips and others, BRS were quick to appreciate that here there was indeed an opportunity to introduce optical disc technology for information products. A number of reasons for our optimism are described below.

Why CD-ROM?

Linked to a personal computer, the compact disc read-only-memory (CD-ROM) offers all the benefits of optical disc technology, with the important new addition that CD-ROM already complies with internationally agreed standards. It is worthwhile here to review the benefits of CD-ROM.

Established standards

Audio reproduction is not complicated by the need to comply with any existing standards, as was the case with the video disc. Moreover the manufacturers of the audio compact disc drives knew that if they were to achieve any penetration of the consumer market, then disc interchangeability was vital. Thus the major manufacturers cooperated fully to define the digitisation standard for the audio compact disc. As a result of their efforts, the success of audio compact disc has exceeded everybody's expectations. Moreover the manufacturers also realised that the compact disc provided an ideal medium for carrying digital information and again have collaborated fully to define a standard for this application. This standard specifies the error checking and correcting techniques which had previously been the proprietary domain of a few specialist suppliers. This aspect of the data application of the disc became the responsibility of the drive manufacturers, since it was catered for by the electronics in the drive.

Thus at once the CD-ROM had resolved all the standards issues that had frustrated the earlier attempts to exploit optical disc technology. Barriers to disc interchangeability still remain, but these will soon be resolved. But tribute must be paid to the manufacturers who, it has been remarked, have already solved 95 per cent of the disc interchangeability problem.

High capacity

The capacity of a CD-ROM is around 540 Mb. To put this figure in perspective it is equivalent to 500 double-sided double density floppy discs, or 200,000 A4 pages, or 20 Kg of lightweight, closely-printed paper.

Data-security

Data on a CD-ROM cannot be destroyed by magnetic fields, and is safe from damage caused by scratches, fingerprints and mishandling. Data is permanent and secure.

Cost effective

Since CD-ROM is a spin-off from the consumer oriented audio industry, the components involved are manufactured to the highest quality using efficient mass production techniques. This ensures that the technology is proven, available, and remarkably cheap. Moreover unlike the situation only a few years ago, the personal computer has become ubiquitous. In short, to the consumer the entry-level price is exceedingly low.

Cost effective for publisher

The CD-ROM is produced by *pressing* from a master disc. For high volumes, this is a very cost effective process. Compare it for example with the production of floppy discs, each one of which must be *serially* recorded from the master copy, and this is only part of the benefit. Compare for example the cost of mailing 20 Kg of paper with that of mailing a single 12 cm disc only 1.2 mm thick. And finally consider the cost of the paper itself.

Flexibility

It is important to appreciate that the CD-ROM is not merely a cheap alternative to traditional methods of information distribution. CD-ROM offers the possibility to combine graphics, voice, music and text to generate totally new kinds of information product. Coupled with sophisticated indexing techniques and the intelligence of a computer, the CD-ROM can be searched instantly to resolve the more complex or obscure query. And since the information is machine readable, it is available for subsequent processing in whatever manner is required.

In summary CD-ROM offers all the well-known advantages of optical disc technology. But in addition many of the earlier problems have been resolved or are no longer relevant: standards are established and almost complete; entry-level cost to the consumer is low.

The omens are good. Why then do we still counsel caution to those who wish to go into full-scale commercial production? The reason is that initial enthusiasm has been generated by those who see CD-ROM as a direct substitute for on-line services. The economic case for CD-ROM in this market is unclear as the following analysis indicates. But more importantly, CD-ROM is an opportunity to escape from the technological constraints of the on-line industry, and to develop products for markets previously unattainable.

Product Economics

The charges described below are presented as typical figures at the time of writing their prime intentions to illustrate the method of analysis. As an example, let us consider a hypothetical database product which attracts the charges to the customer shown below; the on-line wage distribution curve is shown in Figure 16.2.

Vendor charge	£15 per hour
Royalty charge	£15 per hour
Telecoms charge	£9 per hour
TOTAL CHARGE	£39 per hour

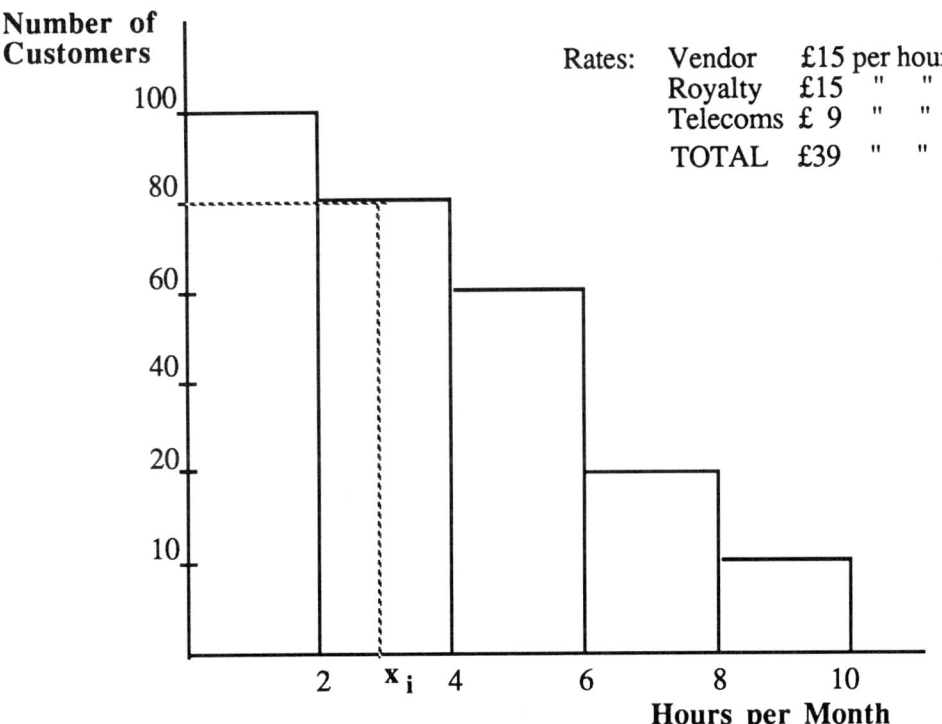

Figure 16.2 On-line usage distribution curve

In deciding to offer the market a CD-ROM alternative, the database producer has to set his subscription price such that the total revenue from his CD-ROM product is greater than his loss of income from his on-line product. Note however that for equilibrium, the database host will respond either by abandoning any on-line service or by increasing his rates to compensate for his own loss of revenue. The starting point for such an analysis is a distribution curve of on-line usage—a typical example of which is illustrated in Figure 16.2.

The total revenue that the database producer would derive from the usage pattern would be:
$15 \sum c_i x_i = 870$ hrs per month $\times £15$ per hour $= £13,050$ per month.

The on-line host also enjoys the same income.

If we assume that the database is around 200 Mb in size, and will be published every quarter, then the fixed costs of production will be £27,000 as shown in Figure 16.3. In addition, each disc pressed from a master will cost £15.00.

So far so good. But now the database producer has to set a subscription price and at this point the problem becomes intractable, as the following discussion indicates.

Let us assume that the subscription price is £S per month. In theory if a customer's monthly usage exceeds S/39 hours, it will pay to transfer to the CD-ROM service. Assume the break-even point is x_{be} at i_{be} (S). Revenue from CD-ROM will be $S \sum c_i$ (where $i > i_{be}$) per month and fixed costs will be £27,000/12 = £2,250 per month, with variable cost being $£15 \sum C_i/3 = £5 \sum C_i$ (where $i > i_{be}$).

Loss of on-line revenue is more difficult to establish and will depend upon the database vendor's response. As already mentioned, either he will increase his connect time charge to compensate for loss of revenue, or he will abandon the on-line service completely. Let us assume he takes this latter approach. Then the loss of on-line revenue to the database producer is the total revenue he previously enjoyed, namely £13,050 per month.

The problem facing the database producer now is to maximise net revenue improvement:
S Maximise $\sum c_i - 2250 - 5 \sum c_i - 13,050$ (where $i > i_{be}$ (S)).

1st disc

Data preparation	£0	(already machine readable)
Data conversion	£2000	(to acceptable input format)
Data indexing	£8000	(full-inverted file)
Data layout	£2000	(High Sierra standard)
Disc mastering	£3000	

2,3,4 discs (updates only)

Data preparation	£0
Data conversion	£500
Data indexing	£2000
Data layout	£500
Disc mastering	£3000

Total annual fixed costs £27,000

Figure 16.3 Annual fixed costs of production

NEW MEDIA: COMMUNICATIONS TECHNOLOGIES FOR THE 1990s

S	$X_{be}(s)$	Net revenue improvement
100	3	21750
200	5	35400
300	8	11250
400	over 10	13050

Figure 16.4

Some sample net revenue improvements are calculated in Figure 16.4. It would appear that the optimal subscription price is around £200 per month, or £2,400 per annum. Of course, this analysis is much too simplistic. Some customers who are deprived of an on-line service will migrate to CD-ROM even though it is more expensive. Others will migrate to CD-ROM because it offers unlimited access. All marketing costs for the CD-ROM product will be borne by the database producer, and these have not been included.

Of all the above assumptions, the last one is probably the most significant because it implies that the database producer will undertake the product marketing. Experience in the on-line industry indicates that this is in general unlikely. The technique in theory could be extended to the situation where an

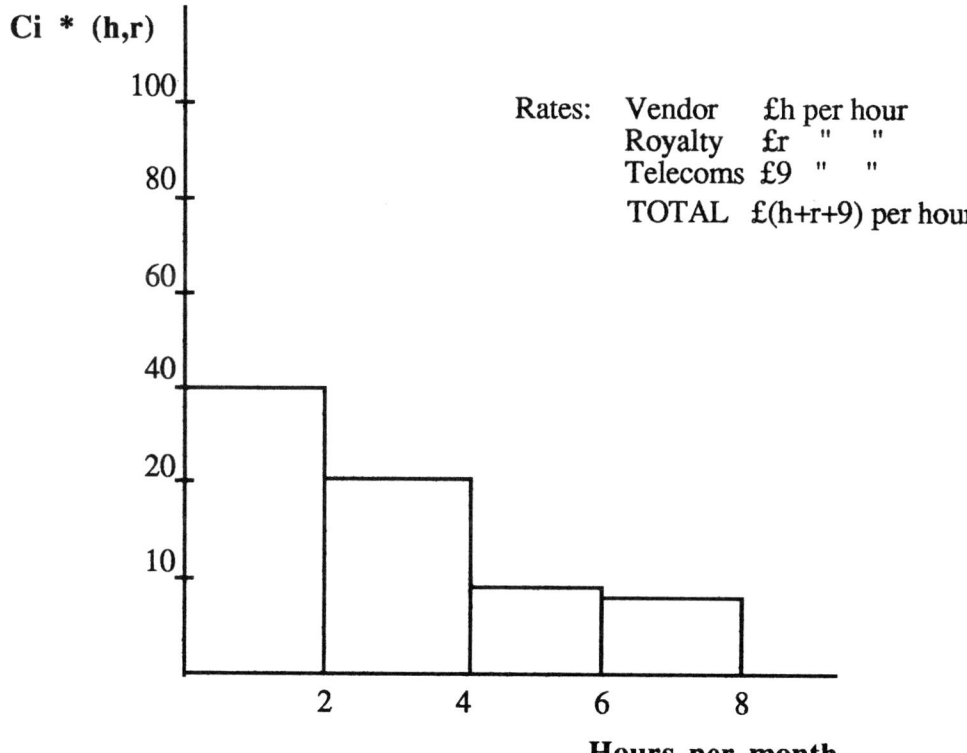

Figure 16.5 General on-line usage distribution curve

on-line service is preserved alongside the CD-ROM service, by the host and producer adjusting their rates as appropriate.

The problem with this formulation is that the new distribution curve c_i is not known. It is only known for current on-line charges, and it is not known for general values of r and h. Its determination would require careful market research.

In summary, if CD-ROM is seen as a substitute for on-line services then the economic case is very unclear. It will depend on the (unknown) shape of the distribution curve and, particularly where the database producer enjoys a high level of royalty payment, it might well be unprofitable to launch a competitive CD-ROM product.

The economic case will also depend, as mentioned above, on who pays the marketing costs. However, the economic justification can be greatly simplified if the host itself becomes involved in CD-ROM production and distribution, and is thus directly compensated for loss of on-line revenue. And this is beginning to happen.

Given the existence of machine readable databases generated for the on-line industry, it is not surprising that the first CD-ROM products have generally sought to replace an on-line database. This is certainly consistent with our own experience, which is described next. But it must be stressed that the interest of BRS is primarily in developing new products for new markets and the examples which will be discussed represent the first step along that path.

Experiences

Many publishers have invited BRS Europe to produce demonstration discs from their data and retained us to assist in subsequent product development. Some of these projects are reviewed below.

In collaboration with the British Library and Philips, we produced one of the first discs in the world which fully demonstrated the immense potential of CD-ROM technology. This was first demonstrated at the Frankfurt Book Fair in October 1985. It is interesting that this disc was produced in nine weeks elapsed time. The disc contained three databases, mostly of a bibliographic nature. It was greeted with great enthusiasm by its intended market, and field trials are still in progress.

In April 1986 we produced two discs for the International Atomic Energy Agency of Vienna. One disc contained the database of the International Nuclear Information Service (INIS); the other disc contained the Agricultural Information Service database (AGRIS). The databases are intended for distribution to developing countries where, because telecommunications facilities are so unreliable, access to an on-line service is not feasible.

For demonstration at the International Newspaper Publishers Association conference and exhibition in Lausanne in October 1986, we produced a disc containing the full-text of all articles appearing in the Financial Times between December 1985 and September 1986. This occupied 180 Mbits on the disc, including indexes. Clearly a single CD-ROM could accommodate at least two years of Financial Times articles.

We have recently been commissioned to produce another disc for the British Library and for Whitakers, the publisher of British Books in Print. This disc was demonstrated at the Frankfurt Book Fair in October 1986 and was the 'largest' disc ever mastered by Philips. Final replication took place in March 1987, when the disc complied with the High Sierra standard and contained MS-DOS and UNIX versions of the software. We will also supply our own device driver routines to support Philips and Hitachi drives.

More recently we have assisted in the production of discs for a number of Italian publishers, concerned with tax law and company information. We are also preparing a demonstration system for a UK publisher of tax law.

At the Frankfurt Book Fair yet again in October 1987 Whitakers **BOOKBANK** was demonstrated. This time the disc was a prototype of the commercial product launched in November. Following extensive field trials, the product equipped with a window interface, offering users a selection of dialogues and a variety of other features.

In addition to assisting publishers with demonstration systems and product development, BRS Europe is also actively involved in more general development work. Our objective is to exploit CD-ROM technology to address new markets and enrich existing ones, some current projects are discussed below.

Human interface

Perhaps the most striking impact of CD-ROM so far has been the attractiveness of the human interface it makes possible. It is in this respect that traditional on-line software differs from CD-ROM retrieval software. On-line software has had to cope with the limited bandwidth of telecommunication links. The PC on the other hand has the advantage of a bit-mapped display, resulting in a very high bandwidth. Allied to colour, the PC therefore offers greatly enhanced capabilities in terms of windowing, menus, pointers, and so on. In terms of software for CD-ROM applications, it is

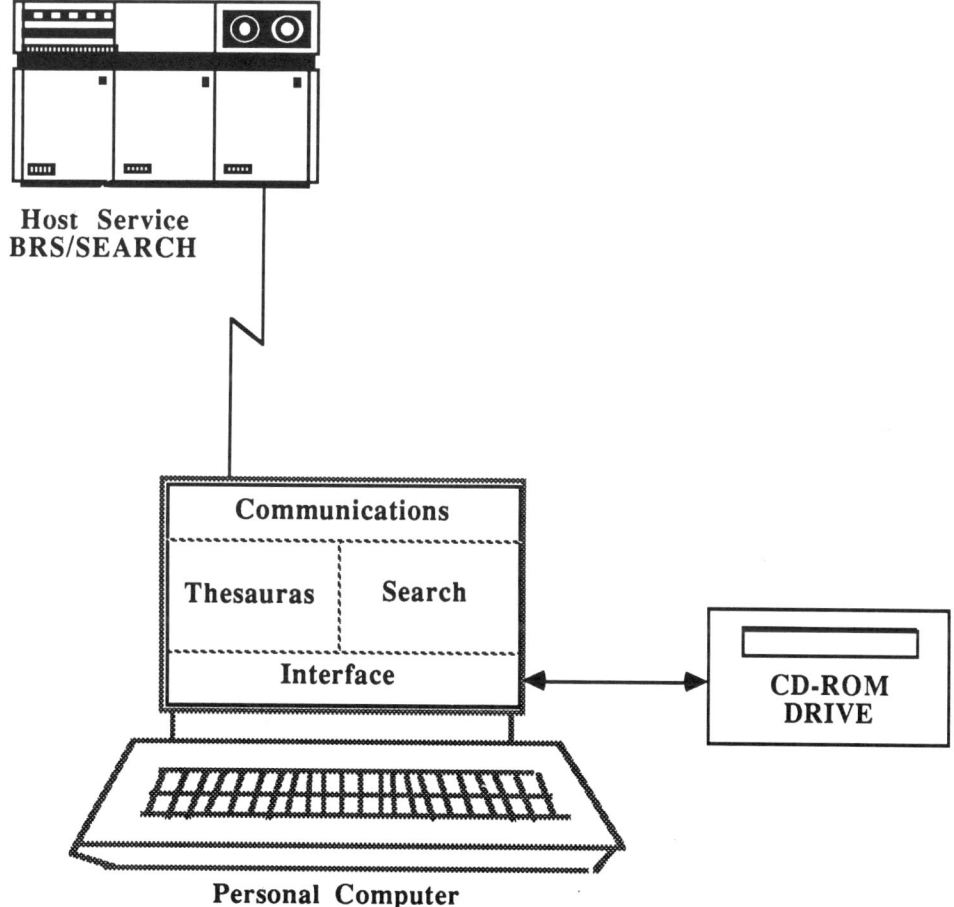

Figure 16.6 On-line access

important that the software provides a set of tools for interface development—it should not attempt to define the ideal interface itself. The ideal interface will depend on the database being accessed and the level of expertise of the user. It is likely that CD-ROM products must offer the user a variety of interfaces, and these should be designed and developed as part of extensive field trials. There is much glitter associated with a number of current commercial CD-ROM product interfaces. However, in our experience these interfaces are only successful at selling the product in the first instance. A user who needs to access the CD-ROM even on a fairly infrequent basis will find such interfaces tedious and ultimately frustrating. Careful guidelines are required.

On-line access

It is obvious that information on CD-ROM will never be as current as that available on-line. It is natural therefore to seek a blend of the technologies which exhibits the best of both. A typical configuration is shown in Figure 16.6, where it is envisaged local CD-ROM data will be complemented by more up-to-date information on a remote host.

The PC will offer the user a uniform dialogue, whether access is to a local CD-ROM or to a remote host. Thus the interface could have all the benefits of windowing, and would formulate commands for onward transmission to the host. In addition, a thesaurus searching aid can be implemented locally, which will complement both on-line and local searching.

Document management

A typical configuration is shown in Figure 16.7. In this example, the CD-ROM contains document images. The index to the documents (inverted file) is maintained on a mainframe host. When a user finds a document of interest, a download command can be used to instruct the PC to extract the document image from the CD-ROM and print it on a local laser printer. Note that the above system is simply an extension to CD-ROM of an existing microfilm management system.

The role of the on-line host

The CD-ROM production process is illustrated below in Figure 16.8. To the point where a tape is shipped to the mastering plant, the functions are generally performed by a so-called system integrator. It is useful to draw a parallel between these functions and the functions which for many years have been performed by the traditional on-line vendor.

The on-line vendor is used to accepting data in all manner of foreign formats. He is used to sitting with the publisher and defining the shape of the database—what fields will be searchable, what fields will be displayable, and so on. And he is used to running the expensive computers which will create the indexes for the database. The on-line vendor has established contacts with the publishing community, and has in place the organisation necessary to manage and execute the above operations professionally. In short, the on-line vendor has years of experience in the role of system integrator. CD-ROM simply represents a new distribution medium; it is merely a very minor diversification in an established and respected service.

Moreover, the on-line host also has a marketing department to help sell the CD-ROM products once they have been manufactured. And as we have already mentioned, his participation in the production and distribution of CD-ROM products greatly simplifies the economic justification.

Retrieval software

The retrieval software is in an intrinsic part of any CD-ROM product. It affects in every way the user's perception of the product. Publishers would probably prefer to remain independent of any

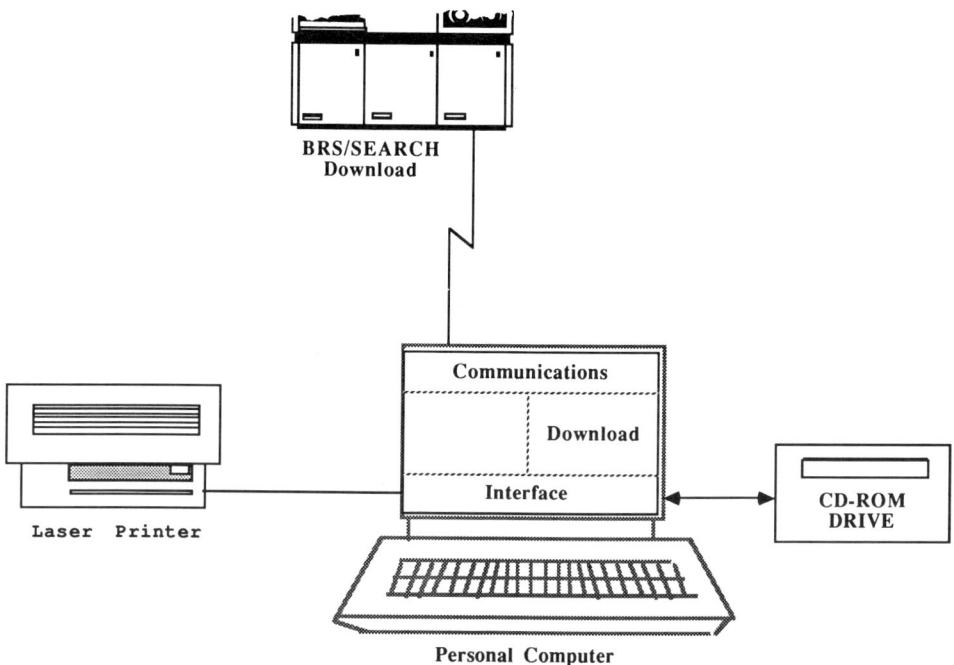

Figure 16.7 Document management

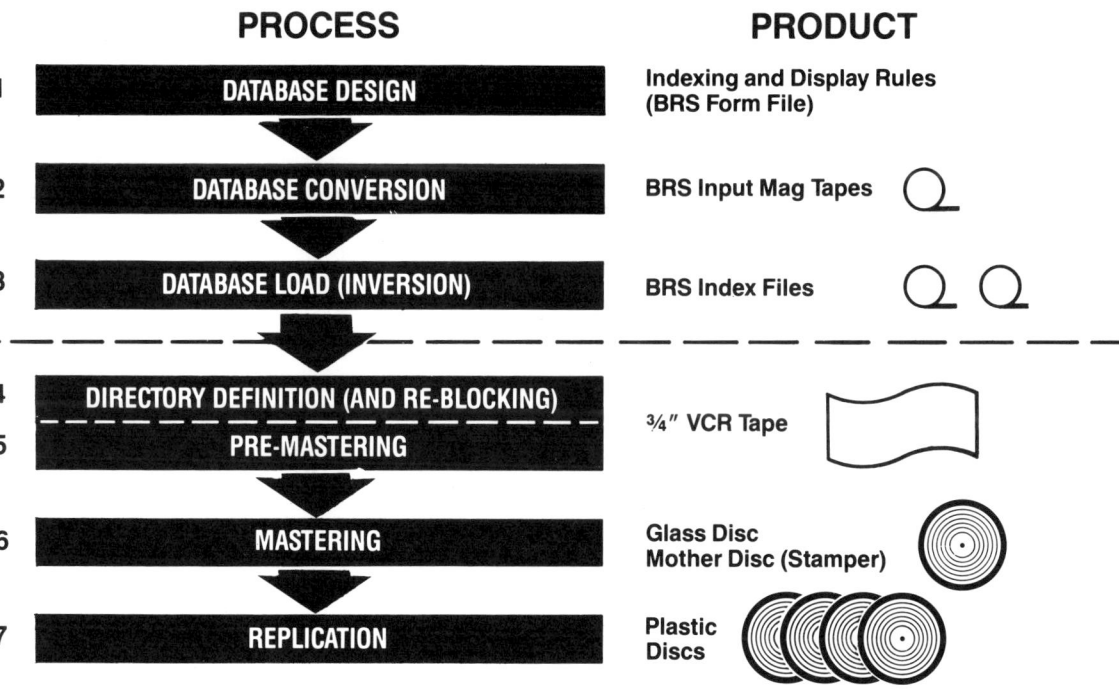

Figure 16.8 CD-ROM production

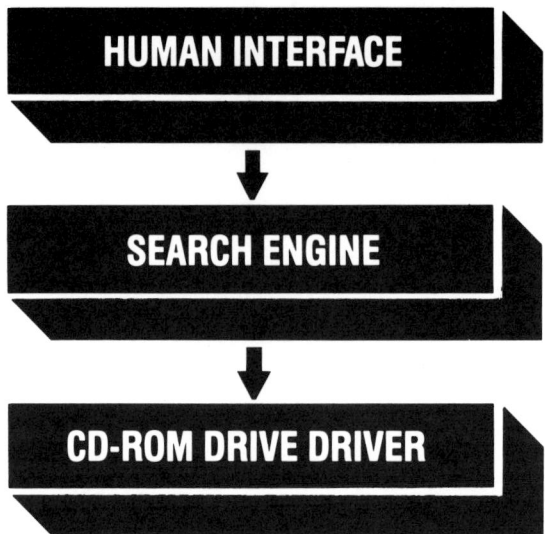

Figure 16.9 Retrieval software

retrieval software, but this is not possible. The disc must contain the indexes (which are defined by the particular retrieval software) and thus the disc can only be accessed by that retrieval software. Publishers must therefore select their retrieval software with care. Indeed we have encouraged publishers to experiment with several software suppliers. It is useful therefore to make some general comments regarding software. There are three main elements to a CD-ROM software product as illustrated in Figure 16.9.

We have already discussed the benefit of bit-mapped display technology in terms of the facilities made available for developing very attractive human interfaces. The search engine receives highly structured queries from the human interface, resolves them by reference to the indexes, and passes the results back to the human interface. It is important to appreciate that a correctly designed retrieval engine makes no concessions to CD-ROM. It is designed to minimise disc seek operations, whether they are CD-ROM seeks or Winchester seeks. The particular characteristics of the CD-ROM drive are catered for by the device driver, not the retrieval engine. This approach has the advantage that advances in the CD-ROM drive technology can be handled by the device driver—there is no need to amend the retrieval engine.

There is a tendency to judge retrieval software only at the level of the human interface and ignore other elements of the retrieval software. We are beginning to hear of the virtues of software specially developed for CD-ROM. Whilst this is understandable, it is also nonsense. What is really being promoted is software which fully exploits the bit-mapped display technology for the human interface. In terms of the human interface, flexibility is the key issue and publishers are recommended to identify this facility in a software product, rather than some exotic design which is superficially appealing but with which real users will become rapidly disenchanted. Publishers should also consider carefully the performance of the retrieval engine element in the software, and the associated CD-ROM device drive. Perhaps above all, they should consider the details of the indexing structure and its impact on how the data can be accessed and at what speed. Full-text applications demand full positional indexing, where the location of every word is indexed precisely—any attempt to string search a CD-ROM database will result in disastrous response times. In other words, the indexing structure must match the application.

Conclusions

In total BRS Europe has produced around 15 databases, or titles, on compact disc. In all cases they have been produced expressly for product development purposes. But the most encouraging aspect of our involvement has been the enthusiasm of the pilot sites for the products. And the most interesting observation is the following—whereas the novelty of the technology wears off in a matter of hours, enthusiasm for the data and the ease with which it can be accessed is enduring.

Although the field trials will go on for some time, it is useful to draw some general conclusions about CD-ROM and its exploitation.

CD-ROM will fulfil its promise. It will offer immense storage capacity at very low cost. It will be an ideal medium for the storage and distribution of vast quantities of information.

The initial market for CD-ROM products will be the traditional user of on-line services who will regard CD-ROM as a cheaper alternative offering comparable response times and a familiar diaglogue.

More interesting perhaps has been the enthusiasm of those prospective users who have never used on-line services, for instance booksellers who have hitherto used microfiche services. To these people, CD-ROM is truly revolutionary. Not only can complex queries be resolved instantly, but the results can be directed to a printer for ordering purposes or to a file for some other processing.

Standards for disc interchangeability are complete. There is a consensus in the industry to ensure that the market is not segmented by a host of incompatible equipment.

The retrieval software is an intrinsic part of the information product. It affects in every way the consumer's perception of the publisher's data. Publishers must therefore select their software, and its supplier, with care.

The software must be flexible enough to enable the publisher to develop products to meet market needs in terms of their human interface. At the same time the software must be fully optimised to guarantee minimum response time and economic load processing.

Above all, the time is ripe for realistic field trials to be undertaken by the publisher. CD-ROM is a truly revolutionary medium and it will have a profound impact, particularly in the professional market-place. Publishers must familiarise themselves with the technology and its potential applications. They must produce sample discs, conduct market trials, and develop the products that their markets require. Demonstration projects are not cheap and will require time and effort. But the investment is necessary for survival. In some sectors, production techniques developed over many centuries will be made obsolete in months.

To conclude on a positive note, and to demonstrate that field trials do lead to successful products, we are proud to report that BRS Europe will participate with J Whitaker and Sons Ltd in producing a CD-ROM version of British Books in Print. This will be one of the first commercial CD-ROM products to be offered on a monthly update basis. A typical configuration is shown in Figure 16.10. It illustrates that CD-ROM not only offers searching capability, but other benefits derived from integration with existing systems.

Once the bookshop customer has selected a book from the CD-ROM database, the next task is to order the book as conveniently as possible. Book details such as ISBN are extracted from the selected record and a record in a preferred format for ordering purposes is stored on the PC. This process is fully automatic and involves no data input by the operator. Similar ordering records will be added to the order file throughout the working day.

Outside normal working hours, a central computer running the ordering system automatically dials-up each bookshop PC and processes the order requests. Each bookshop PC is issued with a list of confirmed orders. Similarly, the CD-ROM database can be used to produce master records for the stock control system. The remote book-ordering system is already in existence and any new facility

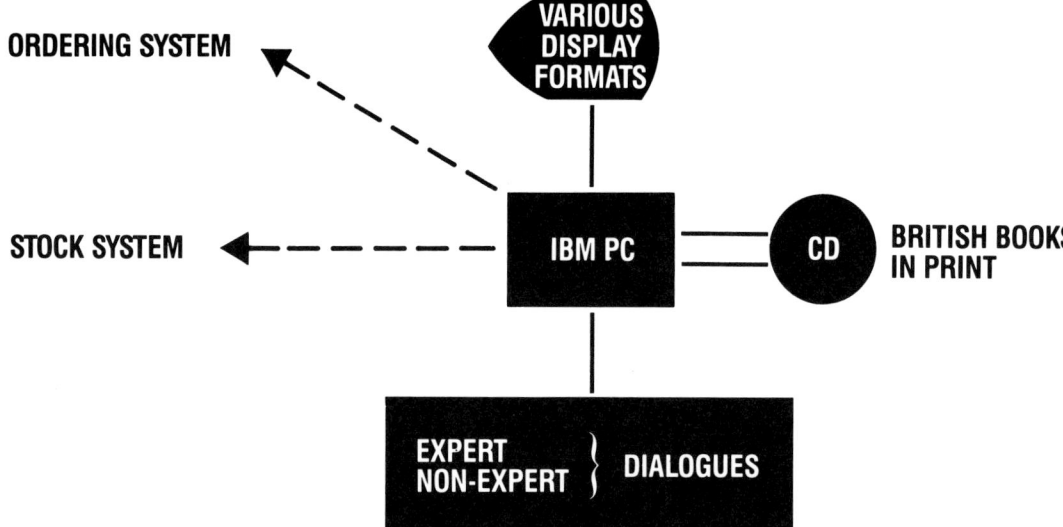

Figure 16.10 Office automation: bookshops

(such as CD-ROM) must integrate with it. This not only requires some bespoke software elements, but also relies heavily on customisation tools.

To UK customers, the CD-ROM version of British Books in Print, known as Whitaker's BOOKBANK, will be offered at an annual subscription rate of £980. For this fee, the customer will be supplied with a new CD-ROM every month incorporating all updates, together with all necessary software, and full documentation. In addition the customer will need to acquire hardware if it is not already available. A CD-ROM drive (and PC-interface card) will cost around £800, for a half-height version which fits very elegantly into a PC chassis. Also it should be noted that CD-ROM drive prices are dropping dramatically—one manufacturer is currently offering them at £450 each. A typical PC configuration to host the system (IBM PC/XT compatible) can cost as little as £2,500, including a built-in CD-ROM drive, colour monitor etc. And even if these prices meet resistance, all hardware can be supplied under a lease agreement.

It is interesting to compare these prices with non-electronic alternatives. The printed version of British Books in Print is sold to UK customers for an annual subscription of £104. There is a single annual edition, comprising 7,000 pages in 4 volumes. A microfiche version is also available for which UK customers pay an annual subscription of £426. A complete copy of the updated database is issued monthly, consisting of over 80 fiches with 270 pages per fiche.

Clearly the CD-ROM alternative is the most expensive media option. But in terms of increased productivity, improved customer service, and increased book sales, the extra cost of CD-ROM is trivial and will be offset almost immediately by its benefits. Happily the market place has recognised this fact and BOOKBANK is being received with enthusiasm by the book trade.

In summary, the technology for CD-ROM is complete. The challenge now is to exploit it by identifying the markets and developing the products. CD-ROM does have a future.

17

CD-I: future applications

Byron M Turner

Will the automobile ever replace the horse? Will the telephone render the telegraph obsolete? Will the printing press ever surpass the hand-written word?

Historically, the most animated discussions concerning new technologies have centered on the topic of future applications. It is not surprising, though, because the beauty of truly revolutionary inventions lies in their challenge to our imagination.

A revolutionary publishing medium

The compact disc interactive technology, or CD-I, is just such a revolutionary development and poses a delightfully formidable challenge. CD-I integrates audio, visual, and text/data functions in a real-time, interactive format that has a vast storage capacity. It not only represents a simple-to-operate, more efficient tool to accomplish established tasks, but also expands the actual possibilities of publishing far beyond the restrictions imposed by traditional means. Because the user need not be bound by limitations of language and literacy, CD-I opens a seemingly boundless frontier of opportunity.

If CD-I is considered in terms of existing technology, you are unlikely to see the larger picture, because CD-I represents an unprecedented synthesis of communications technologies into a new medium. Unleash your imagination, however, and enter the 21st century.

CD-I has the capacity to offer text and audio information in several different languages on the same disc or program. This unique characteristic eliminates the need for the user to understand the particular language in which the material was originally written. The simple addition of visual and audio elements to existing text/data material eliminates even the need for literacy. These are but two ways in which CD-I will change the way publishing is viewed. The main thing to keep in mind is that CD-I offers versatility that will leave no market untouched. Some particular examples will show the extra dimension that CD-I provides over previous technology.

Specific applications

The first and most obvious application is that of extensive information storage and retrieval in a multi-media format. For example, a CD-I encyclopedia might provide the user who accesses the name *Mozart* with, in addition to complete biographical information, pictures of the composer, his birthplace, and architecture and costumes of the period. Mozart's compositions could be heard while the corresponding written scores are simultaneously perused, and animation might depict period dances.

The addition of sound and advanced retrieval capabilities to the traditional methods of presenting text, pictures and diagrams will change training and self-instruction techniques dramatically, both in the classroom and in the home.

Recently, we evaluated how a particular multi-media medical training course for nurses might be adapted to CD-I. The course, a rather good one, involved four video cassettes containing a total of two hours of animation, 465 pages of text and workbook exercises, and 165 35 mm slides. By placing the entire course on a CD-I disc, we could eliminate the need for constant cross-referencing between the different media. We also calculated that the entire course could be presented in three other languages in addition to English, and still fill less than 85 per cent of a single disc.

CD-I educational programs would likewise contain a number of built-in benefits unavailable in existing software. One important example would be the flexibility provided by including different types of teaching aids.

A verbally oriented student might dwell on a verbal explanation of an analysis of certain consumer trends, for example, while a visually oriented student could concentrate on accompanying charts and pictorial displays. The efficiency and effectiveness of each individual's training would thus be maximised.

Catalogues, for both industry and consumer markets, will likewise undergo great changes due to CD-I. Selection of products need no longer by made by merely turning pages. The search process can be need driven, that is, determined by what the user is looking for. If you want to see a selection of men's dress shirts, that is what you will be shown. Perhaps more importantly, for the first time the user can establish parameters for his or her own customised catalogue.

Suppose, for example, you would like to buy a birthday gift for a young boy. Input the appropriate information, including such items as age, sex, hobbies, the amount you wish to spend, and you will instantly create a catalogue of gift options that meet the criteria you have established. Theoretically, it is also possible for the CD-I catalogue shopper to actually complete a purchase transaction electronically, which is certainly an attractive feature to any merchandiser.

The interactive aspect of CD-I again provides another possibility. The digitised information available on each product will, in effect, provide a third dimension. Shoppers will be able to examine any piece of merchandise from any angle, or in any colour, and take a close look at particular details.

This adding of a third visual dimension means a CD-I version of the big luxuriously illustrated book on your living room coffee table will reveal more than just the facades of stately homes and manor houses. With a CD-I program, you will be able to visually stroll around the house and through the sculpture garden. You can even get a bird's eye view, literally, of the scene. There are, of course, extensive applications in such fields as industrial design and architectural design, as well.

A mobile information source

The applications of CD-I will stretch far beyond the home or office and very quickly, no doubt, take to the road. It has been estimated that a map containing every street in the United States would fit on a single disc. This is an impressive statistic but, in light of CD-I's more sophisticated capabilities, almost a Stone Age approach to its use.

This particular point was brought to my attention last year while on holiday with my family. At the end of a rather long day's drive through the French countryside, my wife, whose instincts in these matters are unerring, informed me with a quick glance toward the children in the back seat that I had better find a hotel within the next 30 minutes. The request was simple enough on the surface, but quickly became complicated. What was required was to find a reasonably priced hotel, one which would accept children, within a 30 minute drive. It should also have a reasonably priced restaurant in the immediate vicinity, of course, and be on or near the road we were travelling in order to minimise the possibility of getting lost. If you have ever tried to balance a road map on one knee and a guide book on the other while cross-referencing between the two as the sun sinks slowly from view, you will no doubt be able to imagine the sense of relief a calm, comforting voice from the dashboard CD-I player would have, telling you exactly where to turn off to find the hotel and restaurant you seek.

Before you leave home next time, however, you might wish to employ a do-it-yourself CD-I vacation guide which would allow previewing of attractions and facilities of proposed holidays. To take a sample tour of a possible vacation site, the disc would first reveal a city map grid. The user would choose a location for street photos of the area and steer the narrated tour. Stops inside museums, art galleries or shops, with the ability to browse through the aisles from room to room, could likewise be chosen, as well as similar inside looks at various hotels and restaurant facilities. Favourite stopovers or accommodations could be noted as the tour progressed, so that upon completion of the preview the disc would produce an itinerary coordinated to transportation factors and the user's schedule.

With some idea of CD-I's unique possibilities now in mind, it would be instructive to determine how wide-ranging the system's effect on contemporary lifestyles could be.

Limited to words, numbers, and crude graphics, computer cookbooks have met with little success to date. When it comes to mouth watering dishes, it seems, a picture is worth a great deal more than a thousand words. A CD-I cookbook would provide the same rich visual experience that makes big, glossy cookbooks popular coffee table items, as well as valuable sources of information. Detailed step-by-step illustrations would also make cooking easier. In addition to the added visual dimension, customised cookbooks would be available at the touch of a button, whether the user chooses to organise it by ingredients on hand, favourite foods, special dietary considerations, nutritional goals, or a combination of a number of such factors.

For the sports and health enthusiast, CD-I would be coach and teammate. A new runner, for example, would receive an initial training schedule which would vary according to daily performance information entered by the user. As the user progresses, the program would plan an appropriate diet and suggest special physical challenges to the regimen. Individual medical concerns would also be taken into account.

Electronic games would certainly not escape untouched. From basic word games with random selection which would ensure that no two games were ever alike, to complex exercises of strategy and adventure involving multiple plots, numerous characters and logic puzzles, CD-I will do it all, and with video pictures that would give the games a more realistic look than crude conventional computer graphics.

The arts will likewise be accessible as never before. The synchronisation of words to music will help in learning to play musical instruments and an interactive feature will correct the student musician when an incorrect note or rhythm is played.

A sing-along program would also prove popular for any number of musical styles, from American country and western to Broadway show tunes and rock and roll. With the ability to edit pictures and sound, the user would also be able to produce his or her own music videos, linking sound and/or music to existing pictures.

Music composition, arranging, and even notation of choreography would cross over from the home to educational institutions. So, too, would drawing, painting, and sophisticated graphic design in the visual arts. Once CD-I systems are interfaced with video cassette recorders, personal video editing becomes possible, with professional looking results.

A future of expanding possibilities

These, then, are some areas upon which CD-I will undoubtedly have great impact. Keep in mind, however, that the list of possible applications will continue to grow, for CD-I is the focal point at which all aspects of text, audio, video, and software publishing converge.

CD-I is the electronic printing press of tomorrow, and promises to change our lives as profoundly as Gutenberg's invention of movable type so many years ago. Yet the challenge presented by such new developments remains the same, for the possibilities of the future will only reveal themselves to those with the imagination and the creativity to seek it.

18

CD-ROM and print: partner or competitor

Wolfgang Benscheck

Introduction

In 1979 Verlag Hoppenstedt entered the field of electronic publishing. Information which had already been used as input for print products was now distributed electronically in parallel with the corresponding print products. This paper will try to describe the development of electronic publishing products, the experience to date, also in comparison to corresponding print products, and how CD-ROM fits in this multi-media distribution strategy.

The environment

The main activities of the Hoppenstedt publishing group (a family owned group of companies with 1986 group sales of approximately DM 53 million and 400 employees) are based in Germany. However, Hoppenstedt is also represented by subsidiaries in Austria, Switzerland, Belgium and the Netherlands. Amongst the German based subsidiaries there is one company concentrating on software development and phototype production of print products and one subsidiary, Hoppenstedt Wirtschaftsdatenbank, specialising in the development and distribution of electronic publishing products.

With the traditional print product lines Hoppenstedt mainly serves the German speaking countries and the Benelux area. This means the information published in, and the distribution channels of, these traditional print products are primarily concentrated on these regions.

There are presently three major traditional product lines: financial and stock exchange information, magazines and directories. Products in the area of financial and stock exchange information (company analysis, company reports including balance sheets and profit and loss statements, stock guide, and various charts services) are primarily sold to specialists and executives engaged in capital investment, that is executives in banks, insurance companies, investment trusts.

Magazines, mostly controlled circulation magazines with technical content, are distributed to executives known by name and function, mostly employed in technical divisions, who regularly need information about new products and new product developments.

The third product line, directories, covers mainly buyers guides and directories about top executives in addition to directories containing detailed company information. Users of these directories are mainly top executives as well as division managers in large companies especially in the area of marketing and purchasing. To a certain extent the users are the same as those using the financial and stock exchange information.

Electronic publishing products—first generation

The first electronic publishing product was developed in 1979 based on the information contained in a company directory *Handbuch der Grossunternehmen* (Major Companies in Germany). This directory gives detailed information (company name, address, management, ownership, subsidiaries, sales, employees, equity, industry, etc.) on the 22,000 major companies in Germany.

This first electronic product was organised as an off-line service, providing address and corresponding company information on labels, lists, magnetic tapes and also 8" and $5\frac{1}{4}$" diskettes. The major advantages of this service are high data quality guaranteed through the editorial update procedures for the print product, and high selectivity of data (19 different selection criteria). This off-line service, which was quickly extended to data input from other print products, mainly aims at direct marketing applications, for example in compiling mailing lists, delivering data as input for customer owned marketing databases. Major customers for this product are all companies that use direct marketing in the business-to-business field, such as publishers, advertising agencies, consultancies, banks and leasing companies.

In 1983, the on-line distribution of directory data (company information) through hosts (Data Star, Genios, Dialog) was started in parallel with print products and the electronic off-line service. Applications for the on-line versions are mainly mergers and acquisitions, tracing target groups for sales and marketing purposes, and compiling mailing lists. Major customers for such services include banks, consultancies and information brokers.

Multi-media distribution—experience

A number of observations concerning multi-media distribution of company information are summarised below.

Electronic publishing leads to an improvement of the quality of print products due to the fact that weakness in the information supplied are easily revealed. Firstly, the demand for new information requirements can be identified easily and quickly through electronic publishing applications. Secondly, the use of retrieval and other software in combination with the data quickly shows inconsistencies (wrong coding, different spellings and use of synonyms).

In contrast to print products, electronic publishing (on-line) does not reach the ultimate end user of the information directly. Distribution is through intermediaries, the information specialists.

Electronic publishing has led to a strong internationalisation. New customers in foreign countries were reached, who had not previously been served with print products. Thus, exports stand for 40 per cent of total on-line sales, while exports of print products are only 7 to 8 per cent of total print sales. The number of customers emphasises this trend to internationalisation: 65 per cent of on-line customers are non-German.

Users of electronic publishing products are to a large extent identical with users of the corresponding print products. Electronic publishing activities did not lead to a decline in sales of the corresponding print products. Thus, sales of the print version of *Handbuch der Grossunternehmen* rose by 90 per cent from 1978 to 1986 while the total sales increase including electronic product sales, was 250 per cent. In addition, the number of print copies sold showed a strong increase immediately after the on-line version of this directory was introduced in 1983. While in 1982 approximately 4,600 copies were sold, this figure rose to 5,700 copies in 1986.

Marketing policy for multi-media distribution of information must clearly demonstrate that there is no direct competition between the different media.

Electronic publishing products—second generation

Based on the experience with the parallel usage of data for print and electronic publishing products, there are presently two major new electronic products under development. Both new products, which will be introduced during 1988, deal with company information but are based on corresponding print products of the financial and stock exchange information sector.

The first product provides detailed financial figures (balance sheets, profit and loss statements, etc) of German quoted and unquoted share companies and will be extended later to the major German companies which have to publish annual reports. Data will be distributed using magnetic tape, floppy disc, and via on-line access.

The philosophy of data distribution is as follows: magnetic tape delivery for mainframe applications with major customers (large banks), floppy disc delivery for decentralised PC-applications, and on-line access to satisfy sporadic needs for financial information by customers. In addition to the data there will be a PC-software package available for the floppy disc and the on-line version enabling customers to do financial analysis.

The second product development taking place is a subscription based information system which provides daily stock prices of approximately 3,000 international stocks on a host computer with a back-up history of 600 days. In addition to the on-line access of the stock prices, Hoppenstedt provides a PC-charts-analysis software package which allows the customer to analyse the development of stock prices using a variety of different analysis methods. The product philosophy is to select stocks of interest on the host computer, download the selected data onto the customer's PC and analyse the data on the PC.

Thus the main application is to analyse stock data graphically off-line on the customer's PC. Although the information needed is provided with a small time lag, the purpose of the system is not so much to monitor the share price development realtime; one of the main advantages is the historic information provided. Pricing of the product will be somewhere around DM 7,000 to DM 8,000 per year.

Multi-media distribution—CD-ROM

Based on the experience with multi-media distribution of data as described above the development of CD-ROM has been monitored closely from the very beginning. To evaluate how CD-ROM products could fit in the multi-media distribution strategy practised so far, an initial test product was developed in early 1986.

The input data was again company information from the directory *Handbuch der Grossunternehmen*. The retrieval software used for this test installation was Battelle's Micro-Basis. Problems with access time were found relatively quickly during the four months test period. Company reports naturally do not only consist of textual information, but also contain a certain percentage of numeric data. Thus, problems arose retrieving numeric data. Besides defining technical problems, the test resulted in a detailed catalogue of requirements of a technical and marketing nature for a final CD-ROM product containing company information.

These requirements were defined as follows. Firstly, data input should be the information from two directories *Handbuch der Grossunternehmen* and *Handbuch der Mittelstaendischen Unternehmen*, thus providing information on 43,000 German companies with either sales volume exceeding DM 2 million or a minimum of 20 employees. Secondly, the retrieval software used should also allow searching for and printing of information in a mask driven mode. There should be an English and German language version of the user interface in accordance with the experience of the on-line distribution of data. The application should run under MS-DOS and the most popular CD-players (Hitachi, Sony, Philips). Thirdly, possible applications with the final product should enable a combination of the present off-line service and on-line applications as described before. Comfortable formats for downloading data onto the hard disk should be installed as well as different data interchange formats for further processing of downloaded data with standard PC-software (for example Word Star, DBase, Lotus). Finally, in autumn 1986, a young software company—Dataware 2000, based in Munich and specialising in optical publishing—was found for the development of the software component.

In early 1987, a second test disc was produced together with Dataware 2000, with 90 per cent of the technical specifications installed. One major advantage of the software package developed was the high speed in retrieving both, text and numeric data.

Two problem areas were found for the marketing of the potential CD-ROM product. Firstly, Hoppenstedt has so far been selling information using various media as information carrier. Entering the field of CD-ROM is not just selling information on an additional media, it is more like selling an information application—an information product consisting of various parts: information, software and hardware. Even if hardware were not supplied through Hoppenstedt, a CD-ROM product still requires a different kind of sales organisation and selling method to that used for the products distributed previously.

However, a second, more significant, problem was defined when pricing policy of the CD-ROM product came into discussion. Potential customers for the application described above will actually be the same customers as those for the off-line and on-line applications. Thus, a CD-ROM product in this sense will somehow compete with those electronic services and therefore, pricing has to be thought of carefully, as the following price structure shows:

Print version

Handbuch der Grossunternehmen, DM 530.00 per copy; *Handbuch der Mittelstaendischen Unternehmen*, DM 260.00 per copy.

    36.000 companies = DM 790.00

Off-line service

Price ranges from DM 0.20 to DM 12.50 per company address/report depending on the information scale and output media the customer requires. The average value for magnetic tape and diskette sales amounts to approximately DM 6,500 per order. Orders exceeding DM 5,000 equal 50 per cent of total off-line sales.

    36,000 companies = DM 7,200 to DM 450,000

On-line versions

Connect hour DM 300.00; print charges from DM 0.20 to DM 8.50 per document. Thus, downloading the complete database (including telecommunication charges) of 36,000 companies would result in costs ranging between DM 15,000 and DM 350,000

    36,000 companies = DM 15,000 to DM 350,000

The final pricing decision was made in September 1987. The CD-ROM version will be offered based on a yearly subscription of DM 24,000 including three updates, retrieval software, documentation and one free day introductory training. Hardware will not be included in the price and will not be provided by Hoppenstedt. However, through a sales cooperation between Hoppenstedt and Dataware 2000, Hoppenstedt will be able to help customers with hardware purchases.

Through the pricing policy the potential customer-base was limited to those who use that kind of information intensively. Thus, it will be mainly banks, large consultancies and the top 100 industrial and service companies that will be interested in this product. The potential target group is estimated at around 500 to 600 companies in total.

However, the pricing structure above now allows Hoppenstedt to offer the information in question to those customers who have selective information needs (off-line service), who have sporadic information needs (on-line versions), or who have intensive information needs (CD-ROM), all at a reasonable price level.

The CD-ROM product development will be finalised by the end of 1987 and marketing activities will start in early 1988.

Conclusions

In the past eight years Hoppenstedt's philosophy of distributing company information electronically has been to use information in parallel with corresponding print products. It was up to the marketing strategy to clearly demonstrate that there is no direct competition between the different media (print, electronic off-line, on-line and now CD-ROM) and, indeed, that these are complementary. The selection of one or more of these media is ultimately dependent on the user's requirements.

19
Publishing software on CD-ROM

Eric Coates

Introduction

The CD-ROM medium has been recognised since around 1984 as a publishing medium of great potential. In answer to the question "So what shall we publish on it?", the reply has frequently mentioned the areas of encyclopaedias, databases, directories, volumes of scientific papers, and occasionally that of software. If you use it to publish software, you are making use of its inherent characteristic of being a removable data storage medium for a computer, rather than treating it as a (possibly superior) substitute for an accepted publishing method.

This paper deals with the practical issues associated with distributing software on CD-ROM. It reviews the considerations and problems that apply to any organisation contemplating publishing software by this means. The issues that will be dealt with are technical issues, logistical issues, business issues, and the assorted real life problems that explain the motivation to publish software on CD-ROM and the obstacles on the way to doing it.

Technical issues

The technical issues can be dealt with quite briefly. It is easy to store software on CD-ROM; software is stored as coded computer data in files in exactly the same way as the text comprising a book or a database entry. When the computer reads the CD-ROM disc, it treats it as if it were a magnetic or floppy disc, and so reads in the program using its normal methods. There are of course several categories of software, and all can be stored equally well on the CD-ROM: operating systems, executable programs, and the data they use. In fact it is useful also to regard the documentation for the software as part of the software itself, for the software is useless without it: when you receive software from a manufacturer the documentation needed comes in the same package. There is no difficulty in putting the documentation (a manual or manuals) on the same CD-ROM as the software it belongs to: they simply occupy separate files.

Why publish software on CD-ROM?

Once past the technical question, there are rather more searching questions to ask. Is it desirable to publish software on CD-ROM—why abandon the current methods? Given that it is desirable, is it practical—or is it more trouble than it is worth? And given that it is practical, are the costs involved such as to make the exercise economic? The prospective distributor is not the only one to ask these questions: the recipient of the software will be asking the same—whose interest is served by publishing on CD-ROM, and who will pay the costs?

To find the answer to these questions it is necessary to establish some facts: first of all about the

CD-ROM medium. The CD-ROM itself is physically identical to an audio compact disc and can contain up to 600 megabytes of data, equivalent to 200,000 typed A4 pages. The low reproduction costs per disc give a cost per megabyte (330 typed A4 pages) of about two pence in large quantities. It is reliable, long-lived and, when reproduced in quantity, gives a low distribution cost. The user of the CD-ROM requires a CD reader and a computer. The price of readers (plus interface) is at present between £1,000 to £2,000. The reader transfers data from the disc at 150,000 characters per second, after an initial search-time of (on average) 1.5 seconds. The disc, like any other computer disc, is divided up into files whose contents may be categorised as software or information. The information can be more or less unstructured, for example text, graphics, numeric or binary data; or it may be organised as a database. The computer sees the CD reader as simply another disc drive. In principle, it could be the only storage device on the computer, although in practice this would restrict the computer user in a number of ways.

Those are the characteristics of the technology which is a candidate for solving some of the logistical problems of software publishing. Digital has a major task in ensuring the distribution of its software to its customers, and indeed internally. (The figures following are approximate and for guidance only.) It handles about 130,000 orders per year for software, and has 30,000 software products in inventory, although only about 3,000 are current. In the area of software for the VAX (Digital's mainline computer) there are around 200 'layered' products (layered meaning 'usable in conjunction with each other') in addition to applications software of various sorts. About 10 different types of media are in use for software distribution, from cassette tape to multi-surface hard disc. And as already noted, the package the customer receives contains not only the medium but also one or more items of documentation. Reproducing software on this scale, printing the accompanying documentation, assembling these into products, and putting together products into customer orders is essentially a manufacturing operation, and the area where this is done looks much like a production plant, with warehousing, assembly, consolidation, packing and dispatch areas, to name but a few. Supporting this activity are the production forecasting, production control, and order processing systems that one would expect.

Given an operation of this complexity and magnitude, the potential of publishing on CD-ROM is clear. By putting software on CD-ROM, the media cost is greatly reduced, the variety of media is reduced and so inventory can be lower, fewer media copying machines are needed, and the packaging and transportation costs are lower. Warehouse space is also reduced. Even more significant savings can occur if a large number of pieces of software are put on a single disc, say all the software in a certain category. Unless this is done, a thousand or more different CD-ROMs are needed, some in uneconomically small quantities, quite apart from the fact that a single piece of software only occupies a minute fraction of a CD-ROM. Having many pieces of software on a single CD-ROM not only further simplifies manufacture, assembly, warehousing, and logistics, but eliminates the delay and cost in conveying a new piece of software to a customer site, as it is there on the CD-ROM already. What applies to software applies to documentation also. By putting the manuals onto CD-ROM (probably the same CD-ROM as the software) the same economies can be made for manual production and distribution as for the rest of the software.

Business aspects of software supply

At this point it is appropriate to consider the business aspect of the supply of software from a software manufacturer to its customers. Software is a high value product, which will have cost hundreds of thousands, even millions of dollars to bring to market. The cost of software to the user must therefore represent an appropriate return on this level of investment. Software is not sold outright: the customer purchases a licence to use the software, and optionally other services such as the right to updated versions as they appear. Digital has also recently started to rent some software. Therefore there is, quite apart from the question of production and distribution, the business activities of administering licences and rentals, and software maintenance contracts. This is a major business area.

So far we have seen huge potential benefits in using CD-ROM for software and documentation distribution, that would allow us to get rid of warehouses, slash production administration and transport costs, and give our customers faster delivery of software and at a lower cost. However there are some problems to be solved. This paper will present the problems without attempting to indicate specific solutions, for the solutions for any software publisher will depend on the nature of his business. Nor are the solutions clear-cut: they are in the nature of trade-offs.

It is impossible to distribute software on CD-ROM unless the recipient has a CD reader on his computer. Either the recipient must pay for it, or the supplier must 'give it away'; either as part of the software product, or (if the supplier is a computer manufacturer) as part of the hardware. Who is likely to want to pay? In the short-term the distributor sees the major savings, so is potentially in a good position to fund the distribution of CD readers out of these. The user can expect to benefit from low software prices in the long term, but the anticipation of such benefits will not induce him to invest in a CD reader in the short term. Next, the customer must only be able to use the software he has agreed to pay for, which means that a disc containing many software products or manuals must somehow render inaccessible those which were not ordered. The solution here is to encrypt the software, or at least essential parts of it. This is a technically proven procedure. Finding a foolproof method of conveying unique encryption keys to the appropriate person in a customer organisation is not such a well understood process, and will itself require its own administrative system to be set up. There is also a school of thought that foresees resistance by some customers to paying significant sums of money for something which they already 'have', or for a slip of paper with a code on which will give them access to it.

Finally there are changes needed in the distributor's organisation. Software distribution on CD-ROM cannot possibly replace all of the current publishing methods for software and manuals. There will be many existing computers to which it will be impossible to add a CD-ROM, and for these the traditional distribution methods will have to continue. Distributing manuals on CD-ROM also means that their contents must be read on a computer screen, and this is inappropriate for many environments in which manuals must be used: for example, reading the manual in the bath; or reading a manual while debugging a program at the terminal. This means that in practice the administrative and logistics systems for software publishing on CD-ROM will not replace the existing systems, but will be added to them.

The question then is whether the benefits of reducing the scale of usage of the old systems are sufficient to counterbalance the increased complexity of adding the new systems. It is worth mentioning in this connection that there is no generally accepted software optimised for the function of displaying manuals on-line to a user who wants to use it in the same way as a book. There is in addition a practical difficulty in publishing items of related software on the same disc (for example the 200 items of VAX 'layered' software) namely that almost all of them are under continuous development with releases of new versions planned for a year ahead or more. It would not be practical to distribute a new CD-ROM to all customers immediately a new version of any of the 200 items was ready, so it would be necessary to hold such new versions back for the duration of the publication interval before the next disc (perhaps 8 weeks). This would delay the delivery of new software.

So the decision to be made is not "should we replace our current software publishing method by one based on CD-ROM?", but "is it effective to add CD-ROM distribution to our current methods?" And that is a problem with a lot more variables to be solved.

Experience in publishing on CD-ROM

At this stage it is helpful to look for actual experience in publishing software on CD-ROM. For a fairly obvious reason—that most computers today do not have CD readers on them—nobody is distributing generalised software in this way, whatever they may be planning for the future. But it is open to an organisation publishing a database on CD-ROM to put the database search software on the CD-ROM as well as the data, obviating the need for a separate distribution of the search software. Battelle's Microbasis, which runs on both the VAX and the PC is a case in point, and may be distributed to either of these systems on the same CD-ROM as the database: the Commonwealth Agricultural Bureau's database CD-ROM among others uses this technique. Any database software can be put onto CD-ROM in the same way: the most pressing questions to be asked are not technical questions but ones pertaining to how the software owner is prepared to licence the use of his software. A consequence of licensing will probably be the requirement to collect royalties on copies sold, and so the distributor must evolve a system to do this. So even a technically simple exercise like this requires cooperation between the information owner, the database software provider, the pre-mastering organisation, and the CD-ROM publisher. And of course if the software is to be on the CD-ROM it must be stable and well-proven, because you cannot apply software patches to a CD-ROM.

So although there are major gains which are clearly recognisable, from publishing software on CD-ROM, changing over to this method can in practice be complex. As in so many computer related areas, the main obstacles are problems concerned with the organisation and behaviour of people, rather than technological problems, and it is basically the manager and the organisational expert rather than the technical expert who can unlock the benefits of the technology.

20

The players in the CD industry

Anthony Chandor

This paper covers the two major topics concerning CD-ROM: industry structure and suppliers. The CD-ROM industry is described in terms of the roles played by information providers, integrators, mastering and replication companies, and player manufacturers. Representatives of key suppliers of products and services in each of the industry categories are also described.

CD-ROM industry structure

Information Providers (IPs) are the source of the information which is disseminated on optical storage systems. IPs represent many different types of organisation, and include publishers, libraries, research organisations and commercial organisations. IPs typically use external services in order to create information products for publication on optical storage, rather than invest in in-house expertise.

Integrators are service companies whose function is to produce an integrated information product from the various elements that go to make up an optical storage system. These elements include the raw information provided by the Information Provider, software needed to structure this information into the format needed for optical storage, and the access software designed to work with the database product. Some integrators are also involved in supplying hardware, usually as OEMs for optical storage drive manufacturers, but with some added value provided by the integrator (for example, proprietary error correction/detection sub-systems). Integrators do not normally market product directly to end users, but provide a service to IPs.

Mastering and replication organisations actually manufacture the optical discs themselves. Of particular importance to database publishing are the organisations who produce multiple copies of discs with pre-recorded information. This process is known as mastering, and involves producing a master disc from which replicas can be 'stamped out' in large numbers. Again, these organisations provide a manufacturing service and do not market information products directly to end users.

Player manufacturers make the devices that read and write optical discs. Most players reach end users through intermediaries, integrators or IPs, and are not sold directly to end users by the manufacturer. The aim of most manufacturers is to have their equipment adopted as a standard peripheral to a range of personal computers.

Examples of information providers

Grolier Inc is a long-established US publisher, claiming to be the world's largest publisher of encyclopaedias and reference works, and with over 50 million books a year, is one of the largest hardcover publishers. Its electronic publishing activity is undertaken by Grolier Electronic Publishing one of three companies in its US Reference Group. The main activity in this area has been concentrated on electronic versions of the *Academic American Encyclopedia*, acquired by Grolier in 1981.

The *Academic American Encyclopedia* (AAE) had already been produced in machine readable form for electronic typesetting. Grolier's first investment was to strip out all typesetting codes and produce a database version. This has now been converted to run on 16 major on-line host services. The on-line version of the encyclopaedia contains the full text of the work, but of course excludes any illustrations.

The main lesson learned from the on-line product has been that, while most people using the on-line version of the encyclopaedia ask questions of which 90 per cent are answerable from data in the database, many questions are unanswered or inadequately answered through the limitations of the search methodology. This has led Grolier to research the 'dynamics of curiosity', that is, the way the mind works in seeking out answers to often imperfectly framed questions. This research has resulted in moving away from on-line delivery to off-line products for both 12 inch analogue discs and CD-ROM.

The 12 inch videodisc product suffers from the inherently unintelligent searching facility offered by the home entertainment videodisc player. Grolier see the need to exploit the greater intelligence of a home computer to give a new dimension to computer based searching to the encyclopaedia text. Grolier see CD-ROM as a low cost peripheral to a home computer, enabling software to be developed which will run on the home computer and provide new and powerful ways to seek out information, exploiting their research into the dynamics of curiosity. Grolier has worked with KnowledgeSet on the development of the CD version of AAE, including the development of search software designed to be easy to use by the mass market.

The following are a few examples of CD-ROMs taken from the 150 titles in the marketplace at the end of 1987.

Whitakers publish *Bookbank*, a CD-ROM product containing the full text of British Books in Print, with details of some 470,000 titles and 12,000 publishers. The product uses BRS/SEARCH access software. The annual subscription rate, including monthly updates, is £980.

John Wiley & Sons publish a CD-ROM version of their *Kirk-Othmer Encyclopedia of Chemical Technology* at a price of £580. The product contains the full text of some 1,200 articles and 6,000 tables from the 25-volume print version.

Pergamon Compact Solution produce a CD-ROM version of the 10-volume *International Encyclopaedia of Education* for £1,100, the same price as the print version. Access software, included in the price, is the KRS retrieval system produced by KnowledgeSet, also used in the Grolier system described above.

Oxford University Press has produced a CD-ROM Version of the full *Oxford English Dictionary* on two discs at a price of $1,250. The product is distributed in the US by R R Bowker who themselves produce a number of CD-ROM products including *Books-in-Print PLUS* (sold in the UK for £650) and *Ulrich's PLUS* (£375).

Library Association Publishing produce a CD-ROM version of *LISA* (Library & Information Science Abstracts) marketed by Silver Platter at an annual subscription rate, including, updates of $995. The CD-ROM can be purchased outright (without updates) for $4,995.

Examples of integrators

KnowledgeSet (formerly Activenture) is a small software company based in Monterey, California. The company currently employs 12 staff, and was founded by Garry Kildal, also founder of Digital Research, the company that developed CP/M (a leading operating system for personal computers). KnowledgeSet is best known for its development of the CD-ROM version of Grolier's *Academic American Encyclopedia*. KnowledgeSet is a service company, and does not develop end-user hardware or data products. It describes itself as an optical typesetter, working primarily with publishers on the creation of an electronic publication distributed on optical storage. The company specialises in developing CD-ROM applications, although it has also worked on interactive videodisc systems adapted to text-only displays.

Three levels of services are offered to publishers: data preparation, search and retrieval software and user interface development. The end product is a CD-ROM database, with a file structure including inversion designed to optimise the physical limitations of the CD format, together with information retrieval software tailored to the application and user interface required by the publisher. The data preparation service is designed to take machine readable text and produce as an end product a magnetic tape ready for CD-ROM mastering, including the necessary file structures and indexing.

The company's main investment has been in developing its Knowledge Retrieval System (KRS) software. KnowledgeSet's view is that existing retrieval systems are unsuitable for CD-ROMs, and that there is at present no retrieval software (other than KRS) designed to take advantage of CD-ROM characteristics. The KRS software is matched with the database structure to ensure optimum search access times according to the application, and by separating retrieval logic and output formatting from the way retrieval requests and format commands are input, allows the input formats to be designed for each application. The user interface can be made to look like an existing retrieval system, or can be designed to fit a new category of user (for example for use in the home).

In the UK typical data preparation charges are as follows: conversion of data in printed form to machine-readable form by rekeying or optical scanning, from 60 pence to £1 per 1,000 characters; conversion of a machine-readable file (in typesetter format) to database format, from £30 per megabyte (million characters); file inversion (creation of a full text index), a one-off down payment of £4,000 plus £30 per megabyte.

Data rights are retained by the information provider, and downloading or copying is controlled by software features which make it relatively easy to copy or download small parts to the database but difficult and time consuming to exceed the permitted levels, which are set by the information provider. KnowledgeSet can also provide for data encryption as an additional level of control over copying.

Silver Platter is a service company, providing the services necessary to produce a CD-ROM product from an existing database. The company's development activity is based in London, with marketing operations in London and Boston, Massachusetts. It's customers are publishers or information providers who already control databases in machine readable form. The Silver Platter service is aimed at taking the existing database and producing a CD-ROM version, tailored to fit the medium, and to be used with suitable access software.

The services undertaken by Silver Platter include data conversion into a standard Silver Platter format, the production of indices (inversion) and the customisation of the end-user interface designed to fit the end-user's requirements. A first step is to build a prototype of the CD-ROM version of the database on magnetic tape, which can then be used by the information provider to test the product. Testing takes place before investing in the inversion of the complete database. After inversion Silver Platter produce the necessary mastering tape for production of the CD-ROM by a disc manufacturer such as Philips or Sony.

Silver Platter's own products are an inversion system claimed to be highly efficient at producing the inverted file structures needed for a CD-ROM database, and information retrieval software. The information retrieval software is designed for use on personal computers of sufficient memory size. The software provides the same range of facilities as an on-line full-text search system (such as BRS or Dialog), but tailored to CD-ROM formats. The software offers a 'layered' approach. A tutorial layer provides context related help facilities to the first-time user, which can be by-passed by the more expert user. Silver Platter earns revenue from royalty payments on CD-ROM products produced for information providers.

Typical royalty charges will be of the order of 50 per cent of revenue, part paid as an up-front advance on royalties, used to finance the initial development activity. An alternative method of charging for software is through a software licence agreement. In the UK, Archetype, a leading CD-ROM integrator, changes the following licence fees for inclusion of access software: one to 50 copies, £250 per copy per year; 51 to 100 copies, £200 per copy per year; 101 to 500 copies, £120 per copy per year; over 500 copies, £75 per copy per year.

The product is marketed by the information provider under his own brand image on an annual subscription basis. The subscription includes updates to the database, a CD-ROM player and an IBM compatible personal computer together with the retrieval software. All handling of end users, including the supply of equipment and the delivery of updates is handled by Silver Platter, who in this respect act as an off-line host organisation.

Copyright and downloading issues are addressed by restricting the amount of information which the user can easily store or print at any one run. Encryption is possible, but often the costs of encryption do not justify the resulting value of the protection it affords. Pricing policy is the responsibility of the information provider, but products are generally offered as replacements for high cost on-line services, so the subscription is aimed to undercut the on-line service charge rather than priced on the basis of 'cost plus' on the CD-ROM development costs. Subscription rates for current and planned products by Silver Platter range from $2,000 to $25,000 a year.

Laserdata, based in Cambridge Massachusetts, was founded in 1982 in order to exploit techniques developed at MIT for storage and retrieval of digital data on analogue optical videodiscs. The company is in the business of adding value to OEM optical storage hardware through software and add-on boards to personal computers. It specialises in 12 inch read-only videodisc products, although it is now developing CD-ROM and write-once systems. Laserdata does not sell directly to end users, but acts as a systems house developing products on an OEM basis.

Laserdata will work with a customer to develop a database product. This service includes database inversion, development of retrieval software, and preparation of mastering tapes. Laserdata will also manage the interface with the disc mastering and production activity. A typical charge for database development service would be under $50,000, of which around $8,000 is the disc mastering charge, including delivery of an initial batch of 10 discs.

Laserdata concentrates on two market areas, electronic publishing and interactive videodisc applications. In the electronic publishing market Laserdata see the need for access software to match existing on-line packages in the sophistication of the search parameters, since users will be familiar with on-line searching, and will be unhappy if optical systems offer fewer facilities. Laserdata contrast this approach with the KnowledgeSet philosophy of offering simplified search systems.

The electronic publishing market is seen by Laserdata as being at present a low volume/high price market. Suppliers have to sell a complete product including the hardware, where the hardware cost must be low in comparison with the value and price of the information itself. Again they contrast this approach with the attempt by Grolier and KnowledgeSet to reach a high volume market with a low priced information product. Examples of such low-volume high-priced products include One Source, a financial information CD-ROM issued weekly by Lotus, including 1-2-3 software, for an annual subscription of from $11,000 to $27,000 and Conquest, a marketing intelligence database sold by Donnelley Marketing Information Services at a subscription rate of from $15,000 to $30,000.

Laserdata's view is that the critical factor in a successful CD-ROM product will be to differentiate the product from both print and other electronic publishing formats such as on-line. This will be done for example by adding pictures, graphics, and augmenting the information content, offering more than competitive products.

Mastering and replication

Mastering and replication is a very capital intensive operation, and the activity is undertaken by a few large organisations, such as Philips, Sony, Hitachi, Pioneer and 3M. We will look at one organisation providing these facilities, 3M.

3M (Minnesota Mining and Manufacturing Corporation) is a large US conglomerate, well known for products ranging from Scotch Tape to microfilm. Its primary activity in the field of optical storage is in the production of media, that is the discs themselves, rather than manufacturing players or application development. 3M's investment in media production comes under the Optical Storage Project, based at St Paul, Minnesota, with plant located at Menomonie, Wisconsin. 3M has been involved with optical storage systems since the early 1960s, when it began experimenting with optical recording of television signals on disc. However, these early experiments did not make use of laser beam technology, and 3M's activity remained experimental until the mid-1970s. At that time 3M took the major decision to concentrate its efforts on developing direct-read-after-write (DRAW) write once media, an erasable medium that would operate similarly to magnetic recording, and a simple and high quality replication process for all forms of optical disc. 3M's perception was that data storage applications for the optical disc and short-run replication of video programs would form the main markets for optical recording media, and by 1981 3M had arranged with Thomson–CSF and Philips to offer disc recording and replicating services using 3M materials and processes.

The company's investment in optical storage media production to date amounts to over $100 million, mainly concentrated in plant at its Menomonie site. The main activity today is the replication of short-run videodiscs, but a major expansion of the plant is nearing completion to cater for compact disc (both audio and CD-ROM) replication. Factory space is already earmarked for expansion of write-once and CD-ROM products, and for erasable media. These developments could well double 3M's investment in production facilities to over $200 million. The major current investment at 3M is in CD production, with a new plant coming into production in early 1986. The facility will provide for both audio discs and the CD-ROM data standard. Although 3M are promoting the OROM 5.25 inch format for pre-recorded data, they believe that there is a market for CD-ROM products in the office automation and professional workstation markets.

Typical charges in the UK for CD-ROM production are, for the creation of a pre-master tape, £2,000. The creation of a CD-ROM master disc, including a sample batch of CDs (from 5 to 10): 5-day turnaround, up to £4,500; 10-day turnaround, up to £3,250; 20-day turnaround, up to £2,500. Volume production and packaging are typically: 1–249 copies, £15 per disc; 250 to 499 copies, £13 per disc; 500 to 1,000 copies, £9 per disc; over 1,000 copies, negotiable.

Manufacturers of CD-ROM hardware

A 'buyers guide' to CD-ROM hardware will not be attempted here. One representative company and its offerings is described, giving an indication of this market. Reference Technology Inc (RTI) is based in Boulder, Colorado. It currently employs some 80 staff, and specialises in read-only systems. Key products and services are described below.

Clasix DataDrive 500 is an RTI label on Hitachi CD-ROM drive, together with an RTI interface card which sits inside the personal computer (again an IBM PC, PC XT, PC AT or compatible, with a minimum configuration of 256k RAM and one floppy disc). RTI's added value lies in the interface card. This provides standard PC DOS compatibility, but extends the operating system so that it can exploit the storage capacity and address range of the CD-ROM. The card also contains the RTI error detection/correction processor. The system sells at an end-user price of $1,595 as a complete plug-in-and-play package, with a high volume (1,000+) OEM price of around $1,000.

RTI have a range of software packages developed to provide an interface between user programs and databases held on read-only optical discs. The main product is STA/F Text, a text retrieval package optimised for use with text databases stored on read-only optical discs. The software was derived from a text retrieval package developed by Fulcrum Technologies, a Canadian company, and is similar to the IBM STAIRS system. It was originally developed for the full-text searching of the Westlaw on-line legal database. The system allows for the customisation of the user interface, so that the retrieval language can be tailored to fit the application. The package is licensed with each database, for a fee of $200 per user per title per year.

Access software

Finally, it is imperative that providers of access software packages for CD-ROM applications are mentioned as they are likely to be key players in the industry. Access software will come both from vendors of on-line software packages who develop CD-ROM versions and from integrators developing packages designed specifically for CD-ROM. Among the key suppliers in the area (in addition to the main on-line vendors such as BRS and Dialog) are KnowledgeSet, TMS (marketed by Archetype in the UK), Silver Platter, Status and Assassin.

Development and trends

CD-ROM products are entering the marketplace in increasing numbers. By the end of 1987 some 150 products were on offer. Unit sales figures are hard to come by, but best estimates are that the most popular product has reached unit sales of a few thousands, with most other products achieving unit sales of hundreds rather than thousands.

Interesting products which are now 'real' rather than demonstrations include the Whitaker *Bookbank* database, a CD-ROM version of British Books in Print, and the sale of business oriented products such as the *British Post Office Postcodes* on CD-ROM, marketed by Silver Platter, and *Your Marketing Consultant* (YMC), a marketing information database sold by Knowledge Access. YMC contains maps as well as text, and a number of other CD-ROMs including maps are under development.

A recent analysis of products by subject area shows the following pattern. Adventure, 1; agriculture, 1; catalogues, 1; chemistry, 5; demonstrations, 22; education/research, 4; engineering, 6; finance, 15; general reference data, 18; law/public policy, 10; library, 34; maps, 3; medicine and biology, 16; military/space, 6; natural resources, 5; sociology, 3; software, 1; sound effects 1.

CD-ROM publishers have an increasing choice of service companies able to help with all aspects of CD-ROM publishing: 18 companies in Europe offer consultancy and development services for CD-ROM, including the supply of access software; 16 companies world-wide offer CD-ROM mastering and replication facilities, of which nine are based in Europe. There has been little change in prices quoted for mastering and replication, but turnround times of as few as five days are now offered, making a weekly update feasible. New products coming onto the market include systems which allow in-house production of CD-ROM pre-master tapes. These products are aimed at the large corporate user wishing to develop CD-ROM for internal data distribution.

CD-ROM hardware developments have been largely in the direction of offering units which can be integrated with a personal computer (occupying the space normally used by the floppy disc or hard disc drive). Half height systems (slim-line versions occupying half the standard disc drive slot) have been announced by Philips and Hitachi.

The establishment of the High Sierra protocol as a recognised industry standard, and the availability of operating system software incorporating the standard, has been a significant factor in encouraging the development of new products. Any CD-ROM can now be read by any CD-ROM player.

Future trends

Trends are hard to predict, but the following developments are beginning to point to the future shape of the industry. On-line database vendors are showing an interest in CD-ROM, and this trend could have a dramatic effect, given the huge data resources available to on-line hosts in a format readily transferred to a CD-ROM database format at low additional cost. The growth of hybrid systems, where relatively static historical data is supplied off-line on CD with a 'seamless' interface to an on-line system for current data could be of major importance in the future.

Hardware costs are falling, and the availability of low cost personal computers with built-in CD-ROM drives will encourage the growth of a population of users with the ability to read CD-ROMs.

Access software designed specifically for CD-ROM (rather than being a cut-down version of on-line access software), exploiting the increasing sophistication of personal computers, in particular the advent of high quality graphics and high resolution laser printers, will greatly enhance the ease of use of CD-ROM products and differentiate them even more from on-line products.

21

The outlook for CD publishing

Haines B Gaffner

As momentum builds and the number of announced participants in CD-ROM steadily increases, both sides of the vendor-user equation are asking: How big will it be? Will CD-ROM become as ubiquitous as the floppy disc or remain the special province of well-heeled libraries and corporate data jockeys? The market outlook has changed considerably with the announcement of plans by electronics giants Philips, Sony, Matsushita, and others to market a system called compact disc-interactive (CD-I), an audio/visual-driven enhancement of CD-ROM intended to piggyback on the CD audio system. CD-I opens the door to a true mass consumer market for a CD-computer hybrid—but when, and what must happen first?

Determination of market size

Forecasting future market sizes for new technologies that (like CD-ROM) have yet to establish an appreciable base is a difficult task. As the culmination of a six-month research effort, LINK Resources and InfoTech used a combination of strategies and approaches. An internally consistent model was created through which various price, shipment, and other hypotheses could be tested, beginning with an analysis of current shipment levels, manufacturing capacities, announced products, and the handful of genuine end users now extant. Current and projected markets for published information bases, both electronic and hardcopy/microform, and the degree of CD-ROM eligibility in each were analysed.

An exhaustive survey of library markets to learn how the library community would contribute in its critical role as an early disseminator of CD-ROM-based services to a wider public was conducted. Technology vendors and publishers were discretely polled regarding their plans and expectations.

Previous LINK surveys of business and professional electronic-information-services users were used to bring into focus the nature and sources of demand for CD-ROM databased products in professional productivity markets. Finally, comparisons were made between CD-ROM projections to rates of acceptance experienced by other newly introduced electronic-media technologies and with current and expected sales of other storage media in consumer and professional markets.

Based on such research the expectations for the growth of the overall CD-ROM market between now and 1990 are that a large proportion of CD-ROM expenditures will come not from sales of drives, but from informational discs without which, in the majority of cases, there is no need for the drive.

The sales of content will grow inexorably in proportion to hardware, consituting about 78 per cent of 1990's total CD-ROM systems market. The number of discs manufactured similarly outpaces drives sold, of course; by 1990, close to 18 million CD-ROM discs will have been stamped out for the US market by the likes of Sony, 3M, Philips-Polygram/Dupont, Laservideo, Compuserve founder Jeff Wilkins' Discovery Systems, other record companies such as Warner, and potential new entrants, including R R Donnelly, Kodak, and smaller aspirants.

Professional productivity

The size of the business/professional-information market is estimated to be about $50 billion if you include directly related equipment and services; roughly $15 billion is spent for commercially available information in all formats (inkprint, microform, electronic). LINK market surveys show that more than $2 billion a year is spent in the US for such information delivered in electronic form perhaps $3 billion worldwide.

Drawing on all three market definitions, CD-ROM will find its largest market through the early 1990s in professional applications, although the consumer and education categories will gradually increase their share of the CD-data market. Internal databases (for distribution within a single company) are least dependent upon standardisation. These closed-loop applications will acount for the majority of shipments in the professional-productivity category, especially in the early years. But other professional categories, such as business reference, medical/scientific, engineering/construction/design, and especially law, will account for 90 per cent of the total value of CD-ROM systems sold into professional applications in 1990 because of the high value of the information discs.

Libraries

For libraries, standardisation is especially key. Although their interest in and knowledge of CD-ROM is high, most library management teams are cautious about committing precious acquisitions and equipment funds to a new medium. Out of 510 libraries surveyed, 65 per cent (and over 70 per cent of a special library-expert sample) expected that it would take three or more years for CD-ROM to become a standard information format in their institutions.

Installations at the remaining 30 to 35 per cent will still provide many people outside the corporate/professional world with their initial encounter with CD-ROM systems in the form of electronic card catalogues, indexes to journals and periodicals, and other database services.

Education

Before CD-ROM and CD-I deliver on their instructional and software-management potential, they will have to clear familiar hurdles, such as constrained school budgets, difficulties in digesting and using existing equipment, and a lack of good educational software. However according to Anne Wujcik, director of Talmis Educational Computing (a division of LINK), the costs of software acquisition and software-library management are key problems in the school environment that CD-ROM could address, aided by a re-emergence of interest in centralised mini- or supermicro-based systems to which optical drives could be networked.

The predicted worsening of the 'teacher gap', along with competition between Apple and IBM and among textbook publishers such as SFN, Houghton-Mifflin, and CBS subsidiary Holt, Rinehart, will likely spur development of multimedia CD-ROM and CD-I curriculum materials.

In post-secondary education, state-of-the-art technology is much more important to the major vendors such as DEC, IBM, Apple, and Hewlett-Packard, as well as new contenders like Sun Microsystems or Next Inc. For all of them, CD-ROM will play a role in the development of advanced products for this demanding community.

The consumer market

Grolier's *Electronic Encyclopedia* has served Grolier's long-term and the CD industry's short-term interest by creating some leading-edge consumer awareness. Other consumer pioneers might well enter the market with other reference products that fit the existing home-PC universe, perhaps gracious-living titles such as wines of the world and comprehensive restaurant and hotel databases.

The announcement of CD-I and the evident intent of some major corporations to commit heavily to program creation for this medium has, for now though, changed all bets. The consumer CD-ROM industry (publishers, movie studios, software companies like Microsoft) will soon receive authoring systems needed to create CD-I master discs.

CD-I has the potential to establish a large foothold in the mass market. A previous new technology, the videocassette recorder, has penetrated 44 per cent of US homes, according to LINK's annual home-media survey. Both home computing and CD-I require consumers to change habits and acquire a new mode of playing, learning, working, shopping, and so on.

If the public perceives CD-I as essentially a combination CD audio player and home computer today's advanced households, which are likely to have one or both, might not be enticed to upgrade or duplicate existing devices, and later adopters might not be any more convinced of the urgency of buying anything.

To be assured of success in the wide marketplace, CD-I must offer a completely new kind of experience. It must be an experience that bears returning to frequently. It must compete successfully with existing uses of time, whether entertainment, education, or work-related. The CD-I media system must begin to recreate what both compact audio disc players and VCRs were born with: a pre-existing storehouse of content (records and movies) ready to be transferred to a new medium.

CD-ROM demythologised

Optical publishing is so new and so different that we cannot hope to know exactly how it will proceed. Based on LINK's study of the emerging marketplace and on many years spent observing the (often disappointing) development of new information technologies, the following critique of some early CD myths is offered.

Myth Number One: CD-ROM will democratise information access by bringing the world's knowledge on $2 discs to everyone.

Good information costs good money to provide. If you have good money to spend, CD-ROM should let publishers bring you a great deal of infrequently changing information in a convenient and cost-effective way. If you lack good information now because you can not afford it, CD-ROM will not help you enough to make much difference, at least for a few years.

It is true that many kinds of database information products have become commodities along with computer hardware and software, but those databases that come closest to the ideal of a professional reference package for a given market are very expensive and typically cost thousands of dollars a year.

Most database owners will not risk a nice business conducted at that price level for the promise of a larger mass market for products delivered on lower-cost informational discs for PC users who might become CD-ROM equipped. That is a big part of the chicken-and-egg dilemma for those excited by the possibility of broadening the base of electronic-information users. CD-I's positioning as an alternative audio-entertainment source could be the breakthrough in the home market, but no similar seeding strategy has yet been found in the professional market.

Eventually, entirely new products that are editorially and economically situated between today's books and databases will be developed for CD-ROM and CD-I systems. In the meantime, do not expect a wide variety of creatively packaged, high-volume, mega-reference products.

Myth Number Two: The potential market for CD-ROM is a function of the number of PCs

This myth is based on the premise that CD-ROM is just another storage medium, only with more capacity. CD-ROM is a storage medium on which a regular user cannot store anything without preparing a computer tape, sending the data back and forth across the country or the world, and receiving discs made in a $30-million facility.

For high-volume personal or networked office storage and distribution of large databases to fewer than, say, 25 sites, a different kind of optical storage incompatible $5\frac{1}{4}$-inch write once, read many (WORM) discs (on which the user can write but not erase) make a lot more sense, even though they cost substantially more for both drives and media. There are many other storage media, each of which will find successful niches in the personal-storage market. Defining the CD-ROM market as a subset of the PC market would require the existence of a market category called reference-intensive, or, perhaps, knowledge-based.

Myth Number Three: The companies backing the new CD-I medium are repeating the same mistakes committed in the name of consumer videotex.

Most of the failure of consumer videotex services (the colour-graphics page-oriented kind originated in Britain as Prestel and expounded in the US by the Knight-Ridder and Times-Mirror newspaper giants until both companies shut down their videotex services) can be attributed to the potential customer's need to buy a special terminal or a personal computer (and special communications software even for households with a computer and modem). Another factor is people's tendency not to make sustained use of the medium, especially with pricing based on amount of use.

CD-I is positioned to avoid both pitfalls, in that the concept of CD-I as an interactive viewer that also serves as an alternative audio-entertainment source could be the breakthrough in the home market. Marketed as an enhanced CD record player, it will have a huge base of audio CDs to help justify a purchase, along with more experimental CD-I programs. Also, CD-I and CD-ROM lend themselves to up-front purchase or subscription pricing, so there is no financial damper to regular use.

Myth Number Four: CD-I will make CD-ROM obsolete, at least in the home market.

Many relatively affluent households that are the initial adopters of CD data products already have dual-modular configurations of new media devices: that is, an entertainment centre where the family television, video recorder, and hi-fi system are located and an office with a personal computer and associated peripherals and appliances. With the arrival of popular personal computers with built-in CD-ROM drives, the relatively open and computer-oriented CD-ROM approach will establish itself in productivity applications, whether in the office or the home.

Myth Number Five: CD-I will have no effect on CD-ROM market development.

Most current CD-ROM applications, which involve either text-only on-line databases shifting to local media or text-oriented internal data, will not be directly affected by the prospect of CD-I. Many professional and institutional publishers who have been eyeing CD-ROM as an entry into electronic publishing might opt to wait for CD-I's multimedia emphasis and greater assurance of standardisation.

Educational and library buyers, for whom a single system to handle all electronic-storage needs is highly desirable and standardisation is crucial, will think twice before making major commitments to CD-ROM. In addition, entertainment/education-oriented consumers will sit on their hands *vis-a-vis* consumer CD-ROM products until CD-I is available.

Myth Number Six: The CD-I market will be developed by systems integrators such as Reference Technology, TMS, LaserData, and KnowledgeSet.

Having pioneered the systems and services necessary for optical publishing, in the long run these firms can support only others' marketing efforts. They are small technology-based firms, reliant on others' distribution channels to reach potential CD-ROM users.

If the CD-ROM industry is to burgeon, key horizontal market creators must have names like Lotus, Microsoft, Ashton-Tate, or Borland; broad niches must be served by the likes of West Publishing and its successful technology-based competitor Mead Data Central (law), or by information powerhouses like Dun & Bradstreet and McGraw-Hill. Other well-defined niches will be reached by computer-systems companies like DEC and by some of the thousands of value-added resellers (VARs) serving specific market niches with custom-tailored turnkey applications.

Myth Number Seven: Companies will rush to press CD-ROM discs of databases as a convenient means of disseminating operating information to executive decision-makers.

Those currently marketing CD-ROM believe that, at this stage of the market, companies are reluctant to place on disc data such as internal directories, company-proprietary financial or marketing information, or personnel information. They fear the distributed access could lead to loss of control of such sensitive information.

22

Broadband communications: a global view

David Shorrock

The inexorable trend in communications is towards digital rather than analogue media. In the case of the public switched telecommunications network (the PSTN), as new switches are introduced they are based on digital technology, digitising the speech channels (at 64 Kbps in Europe, 56 Kbps in the US and Japan) and switching a transparent bit stream. But with digital switching comes the realisation that it no longer matters whether the information emanates from a telephone, a facsimilie machine, a computer or terminal. If that digital circuit can be extended out, in the local loop, to the users' premises and the bit stream captured at source, (or in the case of the telephone, digitised in the handset), we have an integrated services digital network, the ISDN.

ISDN assumes digital transmission throughout the network. Fortuitously, the last few years have seen the telecommunications operators rebuilding their trunk networks with high bandwidth optical fibre transmission systems, capable of handling upwards of 140 Mbps. The reason being that fibre is cheaper than copper cable for trunk transmission, and new trunks are necessary because of the almost universal 10 to 15 per cent annual growth in traffic.

However, the more ambitious (commercial) users want more than ISDN: they want trunk bandwidth themselves, for video-conferencing, for high speed data and for interconnecting their PABX and local area networks. So we are seeing the development of *broadband*, which is generally regarded as upwards of 2 Mbps, being installed in the local loop and brought out onto the users' premises.

The commercial user is not alone in his appetite for bandwidth. The consumer market for entertainment, particularly television, is leading many to rethink the role of cable. As new cable systems are installed, the questions being asked are will they just be a medium for television distribution, which perhaps can be done more cheaply by satellite or radio, or will they form the basis of a new telecommunications infrastructure?

Broadband in Germany

During 1988, the Deutsche Bundespost will launch a prototype national broadband network. The project is part of a DM 2.6 billion programme to lay optical fibre through parts of the public network in West Germany. Some 300,000 kilometres of fibre is to be installed both in upgrading trunk capacity between exchanges and, in 29 cities, in the local loops between major corporate customers and the local public exchanges. The prototype network, which will be based on this fibre-optic overlay will offer services such as video-conferencing, video-telephony and high speed data transfer, at data rates of up to 138.24 Mbps. This data rate has been chosen because the Bundespost already had codecs that work at this speed, rather than at the proposed broadband ISDN (B-ISDN) standard of 150 Mbps.

The prototype network will be restricted to 1,000 customers and an investment of DM 110 million, as it is seen as a pre-broadband ISDN service, aimed at stimulating demand for broadband services. The Bundespost estimate that they need to convince between three and five per cent of their subscribers to adopt broadband services in the next 10 years in order to justify their investment in providing the services. The most optimistic Bundespost projections for broadband show demand reaching two million accesses a year by 1995. The most pessimistic show this level of demand only being achieved by the year 2000.

To guarantee an initial group of users, the 35 existing customers of the BIGFON video-telephony service are being transferred to the prototype network. At the end of 1987, the Bundespost claimed that 32 companies had signed contracts for use of the network.

The French broadband switch

The Centre Nationale d'Etudes des Télécommunications (CNET), the research organisation of the French PTT, has developed a switching and multiplexing technique, asynchronous time-division (ATD) multiplexing, which is set to be adopted as an international standard for broadband ISDN.

The importance of ATD is that it allows any combination of data rates to be combined, unlike conventional multiplexing techniques. A prototype ATD switch has been developed at CNET's Lannion laboratory in Brittany under the Prelude project. The prototype switch consists of four matrices each capable of switching 16 inputs and 16 outputs, individually at 4.5 Gbps or at an aggregate bit rate of 9 Gbps. The switch is being used to demonstrate the interconnection of wiring layouts, in the form of local area networks (LANs) of both bus and star configurations, for individual houses. To each LAN is connected a picture phone (64 Kbps voice plus 60 Mbps video), a television (60 Mbps video), broadcast quality sound (two channels of 384 Kbps) and a telephone (64 Kbps). The prototype system uses a CD player, a television camera, videodisc and hi-fi system to simulate the reception of broadcast services.

UK broadband: the jury is still out

In April 1987, the UK Government's Department of Trade and Industry commissioned a study from the PA Consulting Group, to examine how the UK communications infrastructure might develop over the next two decades and the likely effects of this on suppliers, users, and society as a whole. A Discussion Paper of the first phase of the study was published in October 1987 and the final report is expected by mid-1988. According to a senior civil servant on the study Steering Group, the report is for (the Government) Minister's consideration and there is no commitment to publish the findings and recommendations in the public domain.

In establishing the background for the study, the Discussion Paper examines developments in communication technology:

> "The main driving force affecting the future evolution of communications technology
> will be the continued penetration of optical fibre transmission. The introduction of
> fibre-optics would create the potential for considerable amounts of broadband
> communication for all classes of fixed-network subscribers. However, although fibre-
> optic cable is already used in trunk transmission, it will be about 15 years before it
> becomes cheaper than conventional techniques in the local loop. It must be recognised
> that even when this occurs, the installation of fibre optics for purely economic reasons
> . . . would only be carried out for new installations or those which have exceeded their
> useful life."

The Paper then goes on to compare the cost per subscriber for cable networks based on copper and fibre technologies. The conclusion is that the cost of optical fibre will only fall to that for copper by the year 2003.

These initial conclusions (and basic assumptions) are in stark contrast to those elsewhere in the world. In the US for example, Southern Bell is already cabling 500 homes in Orlando, Florida, with optical fibre, and is planning an entirely fibre based network by the year 2000.

As an alternative to cable systems for the domestic distribution of television services in the UK, the Department of Trade and Industry is considering the technology of Microwave Video Distribution System (MVDS), using low power microwave transmitters to distribute multiple TV channels over local areas. In October 1987 the DTI commissioned consultants Touche Ross to conduct a study to examine the technology possibilities for the development of MVDS services in the UK; to consider the forms that domestic broadband MVDS delivery systems might take; to estimate the likely timescale in which equipment might be commercially available; to estimate relative capital costs of MVDS compared with other delivery systems: cable, satellite, additional terrestrial transmission, etc; and to assess the prospects for UK industry.

The results of the study were reported early in 1988.[1] It is understood that the conclusions were that MVDS offers domestic multi-channel TV delivery at lower cost than either cable or satellite, and that low-cost receiving equipment will be available by 1992 or earlier. This is supported by Figure 22.1, which is a comparison between the various media, the costs, availability of the receiving equipment, and availability of frequency spectrum to support additional television channels. For comparison, the table also includes extension to the existing terrestrial UHF transmitter network and the two options for satellite delivery: direct broadcast satellite (DBS) and Astra, the privately funded medium-power satellite (*described by Marcus Bicknell in Chapter 31*).

FREQUENCY	CAPITAL COST PER SUBSCRIBER	NUMBER OF CHANNELS	RECEIVERS AVAILABLE	SPECTRUM AVAILABLE
LOW MVDS (1-6 GHz)	£200-250	Up to 6	1990	By 1992
MEDIUM MVDS (6-20 GHz)	£500	Up to 12	Early 1990s	1990
CABLE	£500-1,600	30+	Now	-
DBS	£300-500	3	1989	Now
ASTRA	£300-500	16	1988	Now
UHF	£30-40	1 or 2	1990	1992

Comparison of Delivery of Additional TV Channels in the UK

Figure 22.1

NEW MEDIA: COMMUNICATIONS TECHNOLOGIES FOR THE 1990S

The European RACE programme

In recognition of both the strategic importance of telecommunications to the commercial and economic health of nations, and to forestall being overtaken by the US and Japanese telecoms industries, the Commission of the European Communities (CEC) is sponsoring the Research in Advanced Communications Technologies in Europe (RACE) programme, at a cost of up to $960 million. In announcing the launch of RACE in 1985, the CEC estimated the annual global sales of telecommunications equipment and services at $40 billion and $240 billion respectively. Further, between 1985 and 1995, spending on telecommunications infrastructure would total $600 billion in Europe alone. In employment terms, and taking into account the information services that are expected to develop to exploit this improved telecommunications infrastructure, it was estimated that by 1990, some 5 million Europeans would be employed in telecoms related industries.

The goal of RACE is the introduction of a Community-wide integrated broadband communication network (IBCN) by 1995. The rationale behind IBCN is that the introduction of digital technology to communication networks changes their economics by means of integration, as voice, data and image are all able to share common switching and transmission media. To achieve these economies of integration substantial investment will be required for laying down the optical transmission media, the development of switching systems, and the development of the terminal equipment and services. All this can only be done on a pan-European scale with the PTTs and the equipment suppliers cooperating towards a single programme.

When first proposed RACE comprised two stages, but during the first Definition Phase, a third was added. The three stages are:

The Definition Phase (1985–86)

Definition of the functional requirements of the network, terminals and services, to assess suitable technologies and to identify areas for further research and development.

The Main Programme (1987–92)

Now referred to as Phase 1, to develop the technology base for the network, trial equipment and services, leading to demonstrations, and to formulate appropriate standards.

Phase 2 (1992–97)

This will depend on the outcome of the previous phase, but will aim to further develop the technology for services beyond 1995.

The concept of the IBCN is illustrated in Figure 22.2 which shows how available data rates will be extended beyond ISDN, from 150 Mbps to 565 Mbps or more. The Figure also illustrates some of the potential services that IBCN will facilitate and the bandwidth that such services will consume.

The current RACE research and development activities leading to a demonstration network include:

Components

The development of low cost components for customer access (TDM multiplexer/demultiplexers, line codecs and so on) using both silicon and gallium arsenide technologies.

Integrated opto-electronics

The development of higher performance and lower cost integrated opto-electronic devices that combine multiple functions on a single chip.

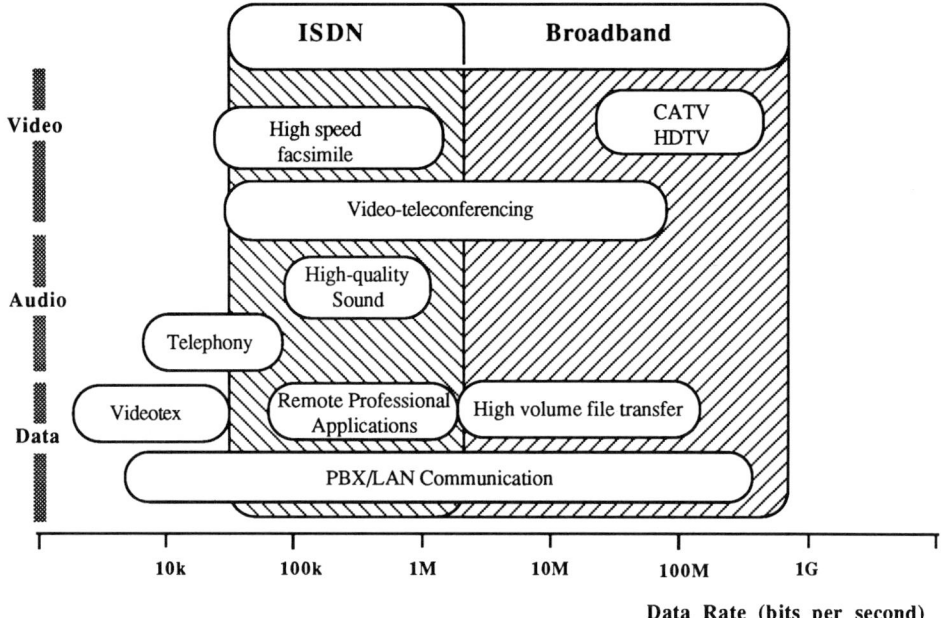

Integrated Broadband Communication Service Data Rates

Figure 22.2

Broadband switching

A review of competing technologies, including optical switching techniques.

High bit-rate links

The development of optical amplifiers and distributed feedback (DFB) lasers for inclusion in the customer loop.

Broadband links

The development of optical fibres and transmission components to support transmissions exceeding 565 Mbps over long distances. Satellite broadband links will also be considered under this heading, but will be operated as intelligent switching machines rather than the passive relay stations currently employed.

Video processing

Extensive work is required in video signal processing and coding techniques to reduce the bandwidth requirements of both present-quality television and enhanced high-definition television (HDTV) transmissions.

Display technologies

Development of small-size high-quality displays which are seen as the key element in encouraging the widespread use of IBC services.

Transnational broadband backbone

In advance of the IBCN, the CEC is sponsoring the development of the transnational broadband backbone (TBB). The TBB is defined as:

> "the concept for a unified approach by the European Telecommunication Operators to provide a main support for the early introduction of advanced telecommunications services by providing high speed digital capabilities beyond national borders in the European Community".

The goal of TBB is to establish, in 1988, a Community-wide digital transmission system of 140 Mbps bandwidth for the initial low volume of broadband services, and to provide a means of interconnection for the various national ISDN schemes. The TBB will be predominantly optical fibre, but satellites will be used to extend the TBB to those areas not served by fibre.

Broadband on a global scale

At an estimated cost of $2 billion, Cable & Wireless are planning a broadband network on a global scale, the Global Digital Highway. The Highway, shown in figure 22.3, will link the world's three main financial markets, North America, Western Europe and the Far East, with an optical fibre network, supplemented by satellites.

The first leg of the network will comprise a pair of transatlantic optical fibre cables, designated PTAT-1 and PTAT-2, to be laid in 1989 and 1991 respectively. Each of the 4,350 mile cables will have a capacity in excess of 3 × 420 Mbps, the equivalent of 17,000 64 Kbps digital telephone channels, or up

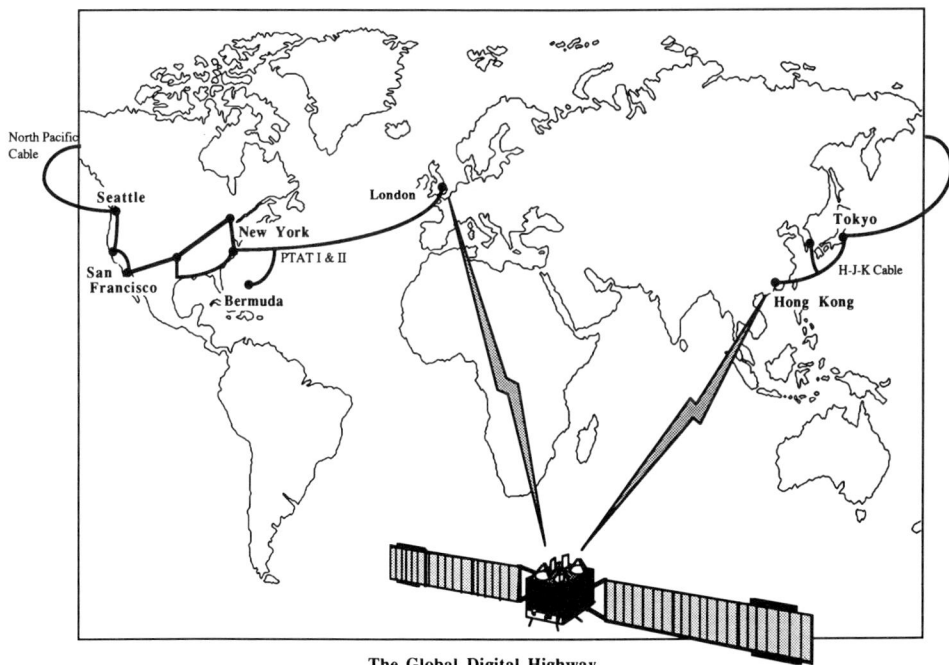

The Global Digital Highway

Figure 22.3

to 85,000 voice channels if circuit multiplication techniques are used. A spur from the cable will go to Bermuda to provide direct connection with the UK and the US. The PTAT cables are estimated to cost in the region of $600 million and are being funded jointly by Cable & Wireless and the Regional Bell Operating Company, Nynex. At the UK end PTAT will connect with the Mercury digital network, while on the US side there will be a coast-to-coast fibre following the tracks and operated as a joint venture with the Missouri-Kansas-Texas railroad.

From the West Coast at Seattle, with a second entry point at Anchorage in Alaska, the Global Digital Highway continues to Japan as the North Pacific Cable. With a capacity of 2×240 Mbps and scheduled for operation in 1990, the North Pacific Cable will be operated as a joint venture with the Pacific Telecom, who have contributed 80% of the estimated $500 million cost. The East Asia link of the Highway, called the "H-J-K Cable System" will connect Tokyo with Hong Kong and South Korea via 4,571 kilometres of fibres, having an initial bandwidth of 140 Mbps. IDC (the Cable & Wireless led consortia including C Itoh, Toyota, Hitachi, Fujitsu, NEC, Merril Lynch and a number of Japanese investment banks) has contributed the $200 million for the Cable System, which is due to be in service in May 1990.

The final link in the Global Digital Highway, returning from the Far East to Eruope, will be via satellite circuits provided between ground stations operated by Mercury Communications in the UK and Cable & Wireless in Hong Kong. These satellite circuits are well established digital routes that are already providing both public and private services. To ensure the availability of these final circuits and to protect against equipment failures, system diversity will be provided by using two satellites and two different earth stations at either end of the link.

Conclusions

In contrast to the IBCN and TBB initiatives sponsored by the Commission of the European Communities (for the laudable reasons of strengthening and developing the European telecommunications industry and infrastructure), the Cable & Wireless Global Digital Highway is based on the commercial judgement of an identified business opportunity. The strength of this judgement is perhaps best indicated by the blue-chip partners, both telecommunications operators and representatives of major manufacturing and service industries, the venture has attracted.

Another contrast between the CEC and the Cable & Wireless approach, is the applications to which the respective networks will be put. IBCN and TBB are very much 'chicken-and-egg' in that the demand for broadband services is unknown but, by creating such capacity, it is hoped that the availability of a broadband network will stimulate and encourage users. Cable & Wireless however have carefully constructed a business case. Some 80 per cent of international transmission capacity is taken up by telephony, which alone is growing at between 10 per cent and 15 per cent per annum. This is before one takes into account the growth in global data transfer spurred by the development of 24-hour financial trading on the markets of New York, London and Tokyo.

On an individual national scale, this contrast between private and public funding of projects is also found. In the US, telecommunication and cable operators identify a business opportunity, pass (or fail) the first test, called 'raising the cash', then go and do it. In Europe, on the whole, the governments and PTTs are funding pilot projects to stimulate demand, as witnessed by the initiatives in Germany and France.

In the UK—with a privatised PTT, token competition and a government that believes in free market forces but still wishes to retain tight regulatory control of telecommunications—the future is uncertain. It is not known what action the UK Government will take on MVDS. Originally, cable television services were seen as a means of bringing broadband services into the home. The cable operators however, are nervous about the cost of laying cable networks in relation to the revenues that would be generated; they see MVDS as an opportunity of providing rapid coverage in advance of digging up the streets. What motivation they would require in order to subsequently replace MVDS with a cable system remains to be seen, unless some regulatory mechanism for enforcement is introduced.

23

Broadband networking
Howard Kleyn

This paper concerns communications networks that are described as broadband. Before considering a limiting definition of a broadband network let us take a brief look at the general concept of broadband transmission and the reason why it has become a hingepoint in telecommunications.

The evolution of broadband

For most of its history the science of telecommunications has been a narrowband activity. It started in the early 19th century as slow-speed telegraphy, which could be described as threadband in this nomenclature. The telegraphic switched service, telex, used a transmission speed of 50 bauds, requiring an effective bandwidth of only 50 Hz. By the end of that century we had the voice equivalent, telephony, which utilised a nominal bandwidth of 4,000 cycles, or 4 KHz in today's terminology. And there matters rested for nearly a century more.

Of course during this period the telecoms carriers, needing bulk transmission facilities, installed increasingly powerful multiplexers which enabled them to allot broadband capacity to a number of users, on a time or phase or frequency division basis. The development of true broadband transmission, however, was left to the broadcasters who first demanded bands of frequencies for high-quality music and then much wider bands for television programmes. (To digress for a moment, this expansion is still continuing. Modern high-definition television, based on a 1,125-line screen format, will need a broadcast bandwidth of some 300 MHz, or about ten times the current requirement).

Turning from broadcasting to commercial telecommunications, the move to broadband transmission and switching that is now occurring primarily in the business world, and shortly in the cable television arena, is based on one fundamental shift in telecoms: from what might be called static information to dynamic information. Static information refers to intelligence that has to be sent only once. All record traffic (telex messages, telegrams, facsimile messages) are carried once, in one direction, by the telecoms carrier. They may go to more than one destination but provided these are error-free no further transmission or switching effort is needed.

Dynamic information, on the other hand, is intelligence that has to be sent more than once—continuously, in fact. A familiar example is the television picture. To enable it to 'move' it has to be constantly updated; this means that a new screenful of information has to be sent to the receiver 50 or 60 times a second. Hence the need for greater capacity, or bandwidth, to handle the information.

This updating principle is now being applied more and more widely to areas of telecommunications that previously rested on the static information principle. Videoconferencing is an obvious illustration (using compression techniques, probably as an interim phase) but continuously updated financial, prospecting and control information is becoming commonplace.

The latter concept, control, whether of a factory process or a guided missile, is the key to latter-day commercial, industrial and military information. In the present Information Technology era it is accepted that 'information is power'; that is rather vague, though. What kind of information and what kind of power are we talking about? The information must surely be (a) accurate, (b) timely, (c) sufficient and (d) pertinent if it is to be of real use. The power is expressed through control; that is, the ability to manipulate people, resources, situations.

The telecoms network operator is in the business of conveying information that is accurate and timely. (Whether it is sufficient and pertinent is for the user to determine.) And bandwidth, or bit-stream in up-to-date parlance, is the key to timeliness. It's the updating principle taken to the nth degree.

A further impetus has been added by the technology itself. The relative annual cost of optical fibre transmission via one 140 M bit/s channel has been calculated by the EEC Analysis & Forecasting Group (known as GAP) to be only one third of the comparable cost of transmission by coaxial cable. The saving is still greater if measured at the two G bit/s level instead of the 4×140 M bit/s level. Therefore, although the cost of multiplexers and repeaters is higher in this technology, the overall cost of transmission of broadband signals is lower today than in the past—and it is still falling.

Broadband networking

Various aspects of broadband communications are covered in other papers. This chapter concentrates on one discipline in broadband communications, namely networking. It was suggested at the outset that a definition of a broadband network would be made; I am going to suggest an arbitrary boundary of two Megabits per second. In other words, any communications network mesh whose links are of this capacity or higher will be termed broadband. Incidentally, the term *broadband* is correct CCITT/CEPT usage. The alternative term, *wideband*, has been used mainly in the UK by the Department of Trade and Industry and British Telecom to describe multi-channel analogue systems; it is not a standard term in the global industry.

A broadband network has one distinguishing feature: it enables a *switched* broadband service to be offered to users. A recent GAP Report has called for a switched bearer service at two M bit/s to be available in Europe in the early 1990s.

At present, as far as is known, only one British manufacturer is offering remotely switchable equipment at two Mbit/s and up, but clearly such equipment will become generally available in due course. Let us examine what it is about the switching that confers such benefits on the broadband applications. That means examining the applications themselves.

Mention has been made of videoconferencing. Alongside it, however, and relatively unnoticed as yet, is a trend towards comprehensive databases. Because of the decreasing cost of digital data storage such databases of text and graphics and image information are proliferating. To be of benefit they will need to be accessed by large numbers of workstations or terminals. It is been forecast, in a European report on the Trans-National Broadband Backbone, that image workstations may account for as much as 10 per cent of personal computer shipments by the early 1990s.

So we have a potential demand for volume data movement (in library applications, etc), image processing and also search facilities, that is to say the ability to 'browse' through a large database. Moreover the search for image quality—three-dimensional high-definition pictures on large screens—adds a multiplier to these trends.

Networks for the future

Where are the stepping stones to the broadband networks of the next decade? One such was Project Universe, which came to the fore in the early 1980s. This was an experimental wide area assembly of

Local Area Networks (LANs) for videotext and teletex applications, linked via satellite and terrestrial paths. It was promoted by Logica, BT, GEC, London's University College and Cambridge University.

On its heels came Alvey, also known as Unison. This was a Logica design supported by the original sponsors less the universities. It is a two Mbit/s net utilising circuit and packet switching based on nodes in London and Manchester. The United States has the Defence Advanced Research Project Agency (DARPA) promoting a broadband packet-switched service based on a single shared three Mbit/s channel. The main application is a videoconferencing facility relayed by a Westar-IV transponder to ten sites.

As stated earlier, the way forward depends on the availability of switching in addition to transmission. Trunk transmission is making its appearance, at any rate in the advanced countries; in the UK it finds its latest expression in Mercury's new all-digital network. Local distribution can be achieved through new ducted cable systems like Mercury's, through cable TV networks or via user premises satcom terminals. Switching at the two Mbit/s level is also beginning to emerge, as noted above, and will undoubtedly spread rapidly over the rest of this decade.

So far, however, switching in public networks has not progressed beyond the bearer level at 64 Kbits/s which is, by definition, not broadband. The next step forward will be switching in the ISDN (Integrated Services Digital Network). The relevant levels are known as Basic Access and Primary Access. The labels refer to a facility known as 2B+D, which utilises 144 Kbit/s capacity, and 30B+D, which utilises 2 Mbit/s capacity. So in this context broadband networking commences at the Primary Access level.

Interpretations of broadband networking

Peering further into the future we can discern a number of interpretations of broadband networking. They include the Global Digital Highway (GDH), Customer-Reconfigurable Bandwidth (CRB) and what I might call the Peacock National Grid (PNG). These are all explained briefly below.

The Global Digital Highway, devised by Cable & Wireless PLC, is a design that interconnects a series of digital broadband paths around the world. It recognises that by the end of the decade broadband international circuits will represent one third of both capacity and revenue on global routes. GDH uses include broker transactions and updating of databases after trading, as stock markets open and close in sequence around the globe, virtual booking offices for airlines and information marketed mainly by press organisations.

The Highway will initially comprise optical fibre submarine cables across the Atlantic and Pacific Oceans, a terrestrial path across the United States and a satellite link from the Far East back to Europe. Spurs to secondary commercial areas such as Bermuda and Eire will be part of the design.

Capacity on the GDH will be offered by lease but the significance of this to broadband networking is what happens in the distribution areas, that is the countries and regions served. It is in these areas that powerful switching techniques will be needed. The PTTs and operating agencies cannot afford to lag behind major private users in this regard or they will increasingly be bypassed by profitable broadband traffic.

As a label, 'customer-reconfigurable bandwidth' is quite a mouthful. Put simply, CRB enables the user to increase the effectiveness of his telecoms consumption while freeing capacity for the Public Telecommunications Operator. This is done by installing an intelligent network multiplexer, under the control of the user, which can, either manually or automatically, adjust the bandwidth devoted to what I have termed dynamic information, one of whose characteristics is its sharply fluctuating volume. The remainder of the bandwidth is devoted to static information (such as file transfer and administrative message traffic) which can be handled during the troughs of the dynamic information flow.

The national grid that I nicknamed the Peacock grid is no more than an idea at the moment. It arises from *Recommendation 15* of the Peacock Report which suggests that national carriers such as Mercury and British Telecom—and any new national licensees—be permitted to act as common carriers for a full range of services, including delivery of television programmes.

If this recommendation were implemented, one or more common carrier cable grids would link local franchises and even extend into their areas of operation. Thus a national broadband network could come into being. This proposal is the subject of a PA report to the Department of Trade & Industry on the Evolution of the UK Communications Infrastructure; its publication is expected shortly.

The role of LANs and MANs

So far this paper has concentrated on national and international considerations in broadband networking. However they must be supplemented by local area networks (LANs) and metropolitan area networks (MANs). LANs may be pactetised (an ungainly verb!) or multi-channel. The latter commonly use a single coaxial cable serving a range of fixed-frequency modems (for data transmission up to 19.2 Kbit/s) and frequency-agile modems (giving a choice of 480 channels in a 24 MHz range or video with telemetry, with up to 48 MHz agility). Frequency-division multiplexing is applied typically to a bandwidth of 450 MHz. Terminals can be easily re-sited by tapping into the cable. Applications include Glasgow Royal Infirmary and airports at Geneva and Zurich.

Perhaps the most ambitious LAN announced so far is the RCA 200 Mbit/s model, planned to go to 500 Mbit/s, for NASA's Johnson Space Center in Houston. The network is designed to serve the planned Space Station and other orbiting platforms from a group of ground stations. It is an example of a 'passive star' configuration, that is no optical switching is performed.

Metropolitan area networks are not fully defined. They are either enhanced versions of LANs or are epitomised by Mercury's City of London network. This is in essence an optical fibre distribution system with the ability to convey broadband services to users over diverse routes from nodes having direct access to national trunks and international satellite paths. It is the first such system to be made available to any of the world's capital cities.

So much for the technical picture. On the regulatory scene the DTI's Technical Working Group on Wideband (Broadband) Cable Systems, known as the Eden Committee, was set up to consider the standards which would be necessary for the design and operation of broadband cable systems. In tackling its work the committee has established a technical policy on optical fibre standards, the introduction of a videophone service, the capacity of ducts and parameters governing stereo sound and digital signalling.

The Eden Committee has submitted more than 100 pages of standards to the British Standards Institution. It has also developed a coordination role between cable operators, equipment manufacturers and broadcasters. In May 1987 the Minister revised the Committee's terms of reference to take account of its expanded coverage.

As regards cable television, the position is that 22 or 23 broadband networks (depending on whether Guildford & West Surrey is counted as one or two) have been licensed, following the grant of the Cable Authority's franchise. Of these, 10 are already operational; they can deliver a range of interactive services in addition to a number of channels of entertainment television. The national operators must obtain a licence, like any other applicant, before they can construct and operate one of these networks. However unless Peacock's Recommendation 15 is adopted they are not permitted to carry television programmes.

It should be borne in mind that while many of the services conveyed by cable television can be delivered direct by satellite or terrestrially via microwave television distribution systems (MTVDS) these are essentially broadcast techniques. They are therefore classed as multipoint one-way systems, whereas broadband switched cable networks are two-way and can accommodate telephony and interactive services of all kinds in addition to multi-channel television.

Trans-national broadband backbone (TBB)

What of the future? In Europe it may depend on the trans-national broadband backbone (TBB) to which reference has already been made in connection with an image workstations forecast. The aim of the TBB is to secure united support from telecom operators for the early introduction of advanced telecom services by providing high-speed international digital capabilities. To achieve this operators are urged to develop a European network of digital highways based on enhancement of existing and planned capacity. The elements of the TBB are services, demand, switching, transmission and cost.

Services

The consultant's report notes that no standardised services at higher than 64 Kbit/s exist and that videoconferencing is the emerging requirement. It recommends that bearer services be introduced in the order 2, 140, 8, 34 and 565 Mbit/s, with a full switched service at all rates by 1995.

Demand

A couple of ISDN-based scenarios have been mooted but much market research is clearly needed.

Switching

Circuit switching with CCITT No 7 signalling is recommended; no technology is yet available for integrated narrowband/broadband networking.

Transmission

Digital mode over optical fibre is recommended, supplemented by satellite paths.

Cost

Estimated to lie between £220 and £360 million.

It is evident that broadband networking is still in embryo, let alone in the cot. The necessary transmission paths are largely in place but the vital switching has yet to be developed. Local distribution is even less imminent. It remains to be seen whether the twin forces of ISDN and cable TV will combine to bring broadband networking firmly into the national telecoms infrastructure.

24

Perspectives on coding, switching and transmission

J R Howard

The evolution of telecommunications networks is shaped by two strongly related areas: services and technology. Until recently services have been limited by technology to a very basic set and the main thrusts of technology have been towards minimising cost and maximising quality of these services.

Technology is advancing in two contrasting areas: to increase the bearer capacities of telecommunication networks and to reduce the bandwidth demanded by services. These advances have opened the way to a new generation of services, by reducing their network requirements, whilst simultaneously making much more capacity available. This chapter examines the potential services that will make demands upon broadband networks, the network components that will support these services and the technologies that create or modify the broadband network opportunities.

Broadband services

Some of the services that are anticipated over the next ten years are summarised in Figure 24.1. In this figure, the services are distinguished by the data rate applicable to them. It can be seen that whilst today's services are based upon the ISDN 64 Kbit/s data rate, future services under consideration range from 2 Mbit/s to 140 Mbit/s—an increase factor of approximately 1,000. These new services are referred to a broadband services.

It is widely assumed that the dimensions and connectivity ratios of future networks and network components will remain roughly as today. Consequently, the increased performance factor of approximately 1,000 will manifest itself in every one of these components.

Network components

A simple network picture is shown in Figure 24.2. The performance characteristics of the various components are summarised below and compared with current networks.

Subscriber to exchange link

In the current network, ISDN standards are settling on two data rates for this link: *basic rate*—running at 144 Kbit/s supporting two 64 Kbit/s data channels and one 16 Kbit/s signalling channel (2B+D); *primary rate*—running at 2 Mbit/s (in Europe) supporting up to thirty 64 Kbit/s data channels and one 64 Kbit/s signalling channel, multiplexed onto one physical link.

For the future broadband network, a typical aggregate link speed might, for example, be 565 Mbit/s supporting 3 channels at 140 Mbit/s (broadcast HDTV), 4 channels at 34 Mbit/s (videophone), and a number of channels at 2 Mbit/s or below (ISDN services).

LIST OF POSSIBLE VIDEO SERVICES IN THE IBCN AND THEIR REQUIREMENTS

VIDEO SERVICE	APPLICATION	BIT RATE
Video telephony	Person to Person communications	64 kbit/s (384 kbit/s)
Video conferencing	Group to group communications - need for high resolution graphics capability, high quality audio, encryption for security	(64 kbit/s) 384 kbit/s - 2 Mbit/s
Surveillance	Security; remote control; traffic monitoring	(64 kbit/s) 384 kbit/s - 15 Mbit/s
CCTV	Closed user group broadcasting; remote lecturing	15 Mbit/s 34 Mbit/s (70 Mbit/s)
Video Library	Entertainment; remote education and training transmitted on request; interactive video games	34 Mbit/s - 70 Mbit/s
CATV Distribution	Entertainment including pay-to-view	34 Mbit/s - 70 Mbit/s
Enhanced Definition and High Definition Television	Distribution	140 Mbit/s - 280 Mbit/s

Figure 24.1

It should be noted that the current state of the art in installed trunk transmission systems is 565 Mbit/s with 2.4 Gbit/s an imminent prospect. Clearly the technology improvement will have to be significant in order to provide a cost effective subscriber access.

Switching

At present, most digital switches employ a time division multiplex technique in which incoming circuits at 64 Kbit/s are multiplexed up to 2 Mbit/s. Internal connection within digital switches

SIMPLIFIED DIAGRAM OF NETWORK COMPONENTS

Network
Termination
Exchange
Local loop
Long haul
transmission

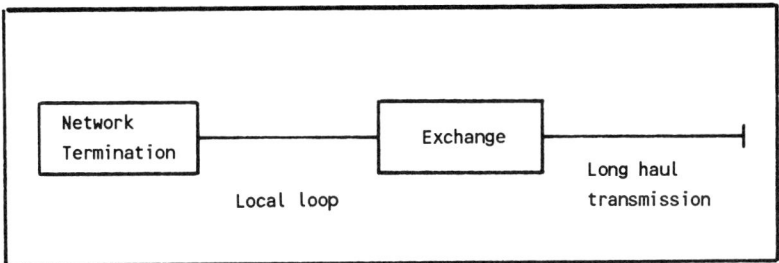

Figure 24.2

IMPROVEMENTS IN PERFORMANCE OF LONG HAUL SYSTEMS

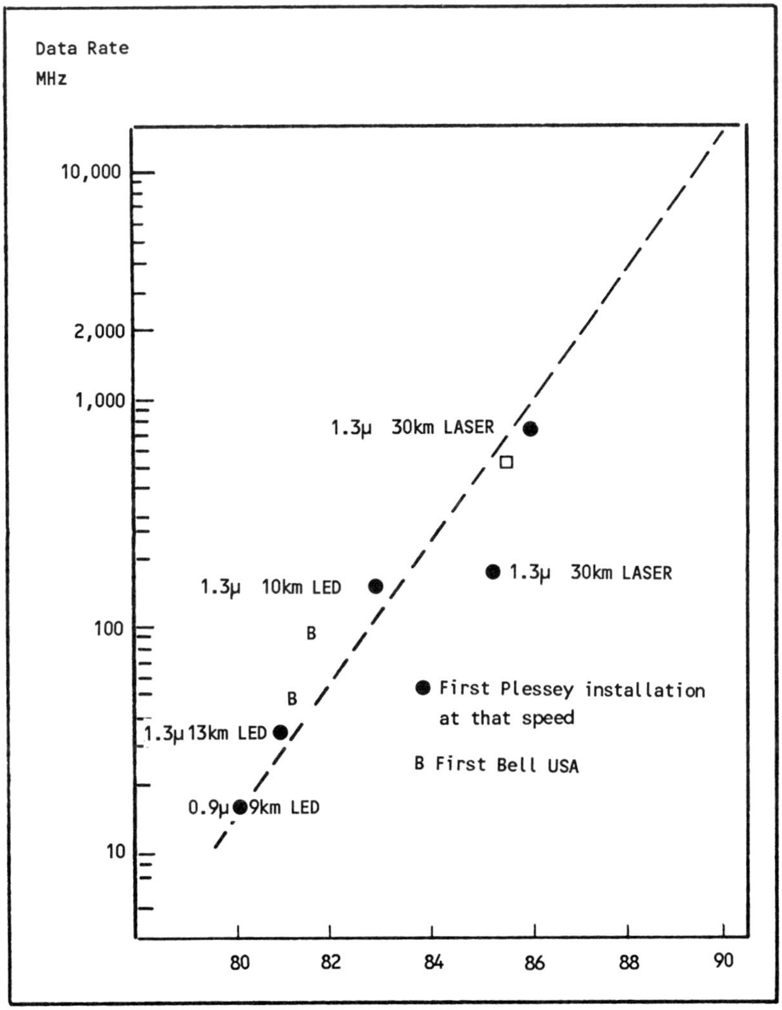

Data Rate
MHz
10,000
2,000
1,000
100
10
1.3μ 30km LASER
1.3μ 30km LASER
1.3μ 10km LED
B
B
1.3μ 13km LED
0.9μ 9km LED
First Plessey installation
at that speed
B First Bell USA
80
82
84
86
88
90

Figure 24.3

typically run at speeds of 8 Mbit/s. The technology of time-switching is constrained by a number of factors including memory access times, logic complexity and timing problems associated with interconnect. In order to switch circuits at rates of 34 Mbit/s and 140 Mbit/s, either a dramatic improvement in technology or a rethink of the internal architecture of the switch is required.

Long-haul transmission

In this field there has been a steady improvement in the performance of installed systems, illustrated by Figure 24.3, which shows a doubling in performance every two years. Clearly, with the introduction of broadband services, this performance improvement will have to accelerate in order to keep up with the demand.

Review of technology

Integrated circuits

There are a number of expected improvements to existing technologies which will be achieved largely through refinements in the production process, which progressively make the feature size smaller, resulting in improved speed : power ratios. Additionally, new materials are beginning to emerge which promise even higher performance. Three of the better known technologies are silicon CMOS, silicon ECL and gallium arsenide ECL. The relationship between clock frequency and power per equivalent gate is shown in Figure 24.4. It can be seen that with diminishing dimensions, maximum

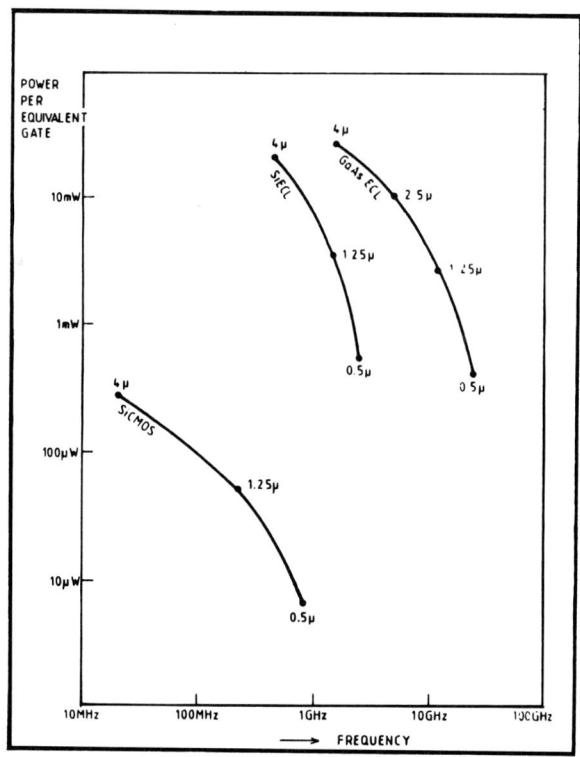

Figure 24.4

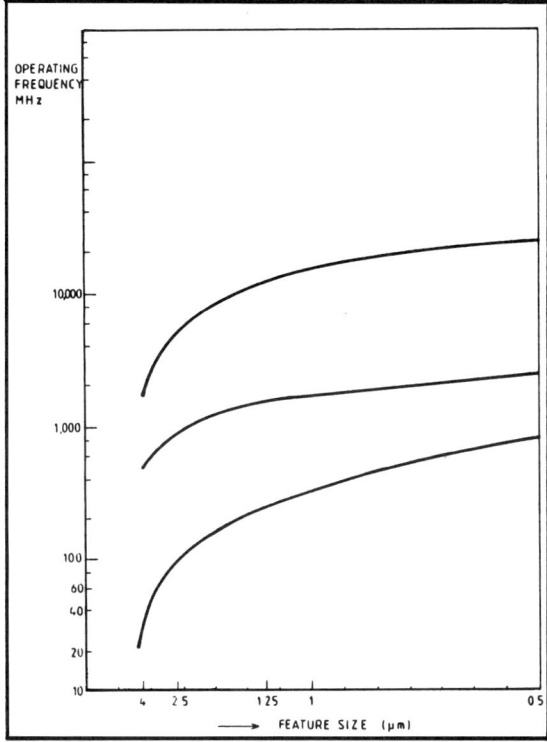

FEATURE SIZE versus OPERATING FREQUENCY

Figure 24.5

frequency increases and power dissipation decreases. It can also be seen that, in order to achieve the higher frequencies of operation, the ECL transistor configuration is needed, which can consume up to one thousand times as much power as CMOS. Figure 24.5 shows the relationship between feature size and operating frequency.

Two conclusions can be drawn. For *transmission systems* involving very high operating frequencies but relatively simple logic, gallium arsenide ECL offers the potential of frequencies of greater than 10 Gbit/s. For *switching systems*, CMOS will certainly cater for space switching of channels up to 140 Mbit/s and with suitable design and system architecture can probably handle the time-switching of 34 Mbit/s channels. These results were confirmed in a number of projects within the RACE Definition Phase.

Integrated opto-electronics

The advances that have been made in optical communication in recent years are shown in Figure 24.3. This suggests that transmission rates of many gigabit/s may be available before 1990. Typical components required to enable this performance to be achieved are zero dispersion fibres at the 1.55 μm low-loss window, directly modulated injection lasers to 20 GHz, 100 GHz detectors and high speed logic to 9 GHz.

An increasingly important technique is to use wavelength multiplexing in which light signals from a number of optical sources are separately modulated and optically combined into one fibre. Research is being carried out to improve the sensitivity of receivers by the use of coherent techniques. These techniques can potentially, allow hundreds of channels of up to 1 Gbit/s to be multiplexed down the same fibre.

Research is being undertaken into the possibilities of optical switching. One technique, for example, uses electrically controlled changes to the refractive index of a waveguide to alter the optical lightpath direction. These systems all have the characteristic that once a path has been selected, the switched

VIDEO QUALITY CLASSES V.S. COMPRESSION

Video quality class	Magnitude of bit rate for coding (Mbit/s)		Examples of application of service
	uncompressed	compressed	
High-definition TV-quality	~1000	~800	TV-signal transmission studio-studio
		~400	Very high quality videocom.
		120-140	TV-distrib. in the local loop
Normal TV quality	216	135	TV-signal transm. studio-studio
			Normal TV qual. video communication
		34-70	TV distrib. in the local loop (end user)
			Normal TV qual. video communication
Low TV quality	~40	1.920 0.384	Video Communication
Very low TV quality	~5	0.064	Video Telephony

Figure 24.6

bandwidth is almost unlimited. The consequence of this is that it will be possible to build very high capacity space switches for use either at the heart of switching systems or as transmission flexibility points.

Coding technology

Whilst research is continuing to expand the information carrying capacity of transmission and switching systems, there is a considerable body of research aimed at reducing the bit-rates that are required to be sent, so that those systems can be used more efficiently. Particular attention is being given to bandwidth reduction of video signals, because it is known that there is considerable redundancy in those signals which can be exploited by suitable coding algorithms and technology. It is hoped that shelves of codec equipment in today's technology may be reduced to a few chips within a few years.

Figure 24.6 shows the main service classes, their uncompressed data rates and examples of potential compressed data rates. Clearly, as the compression ratio increases, the quality of the received signal can deteriorate. Also, as the compression ratio increases, the complexity, and therefore the cost, of the logic increases.

Impacts of technology

It is interesting to speculate on how the future performance of video coding will affect the network. Current video codec systems are very expensive, being largely based upon discrete or small-scale integration. The RACE program intends to develop chip sets probably based upon $1\,\mu$ CMOS which will significantly reduce the cost of codecs. If a truly cheap codec can be produced which achieves an 8 : 1 data reduction and maintains acceptable quality, it is likely that TV signals will be distributed at the lowest possible rate, consistent with quality, taking the pressure off the local transmission systems. If, however, the cost of the codec remains high, it might be economic to share the use of codecs at the local exchange. This would have the effect of increasing the data rate in the local loop which clearly has an effect on local loop cost.

Certainly in the long-haul systems, it seems quite likely that codecs could be used to reduce the transmission data rate requirements. One implication of this sort of solution is that the network must be aware of the use being made of individual links and, furthermore, must know the standard that is being employed. These are complexities that the current network does not have to deal with.

For the immediate future, there seems no doubt that broadcast television services will be switched at some network node and a limited set of channels distributed to each subscriber on a demand basis. This is because of the fundamental transmission constraints in the local loop. However, if sufficient capacity is available in the future, it may become cost effective to distribute all channels to every subscriber and select channels at the subscriber premises. This would avoid the need for a separate switch in the local exchange. Clearly, this is one interesting trade-off which must be evaluated in the future.

Conclusions

The anticipated trends in services have been outlined and the technologies that are required to support them have been briefly touched upon. There is every confidence that the technology capabilities will continue to satisfy the demand and enable future broadband networks to be assembled.

The European RACE initiative is exploring the techno-economic options available within the integrated broadband communications network of the future. As indicated, there are a number of technical alternatives available which affect the cost and performance of network components. The alternatives under consideration can imply either the use of different network components or varying the location of components in the network. It has been recognised that, because of the rapid movement of technology, the optimum network solution will vary in time. Great care will however have to be taken in planning the correct evolutionary path for broadband networks.

25

On-demand interactive video on British Telecom's Switched-Star network

Gordon W Kerr

Introduction

The British Telecom video library concept, as part of the service package of the BT Switched-Star Cable system[1], can be considered to be a fundamentally new service for cable-TV, and one which could eventually contribute considerably to the viability of advanced broadband local networks. Perhaps for the first time on a commercial network, a customer using this facility will be able to break out from the stream of pre-scheduled programmes and choose to view what he wishes when he wishes. There will be access to a large catalogue of entertainment and instructional programmes, through the controls on the keypad, and in addition, the customer will also be able to interact with what he chooses to view. The video library, however, will not just be an on-demand film service; it will be able to support a wide range of interactive video services—a field of very active innovation[2]. The network allows for many customers to share a common central interactive video centre and this offers many opportunities for new and attractive services which few customers will be able to afford to own personally. The range of programmes the library can support includes: films and documentaries, pop videos, video encyclopaedia, DIY instructional material, interactive educational video (programmed learning), and home shopping catalogues, including active demonstration.

One library module is designed to serve up to 250 simultaneous sessions and it is estimated that this should be adequate for a sector of the cable-TV network with between 10,000 and 20,000 customers connected.

This paper describes the video library service and outlines the structure of the system.

User overview

Library session overview

A customer asks for library service by pressing the LIBRARY key on the keypad. An on-demand video channel is then allocated to the customer for the session, and he is switched through to the library. The customer then enters a text dialogue to request a programme and to exchange password information. The library computer switches the allocated video source to the allocated video channel and the customer's programme begins. Customer keystrokes are directed through the network control system to the library computer, which directs them to the appropriate video playing hardware (video source).

Once a session has started, the customer can use the keys on the keypad to control his viewing. For standard catalogue programmes, like films, that control can be simply stopping the programme, going over a part of the programme again, or skipping over another part. For more sophisticated items, the customer may wish to learn interactively from a DIY 'How to' programme, or study something from a video encyclopaedia: the level of control the customer is offered in these programmes is fixed by the programme creator, not the video library itself.

To end the session, the customer presses the FINISH key. This causes the session to be cleared down at once. Alternatively, if the library detects the logical end of a programme, this can also cause the session to be cleared down. Charging is performed by the library computer on a basis of a charge for accessing the programme plus a charge related to the time spent viewing the programme; billing packets, which include a credit for the programme rights owner, are sent to the system administration computer.[1]

If the customer chooses an item like a film or elementary information programme, he will be switched to standardised video playing hardware (a standard video source), which has a pre-defined user interface for all programmes played on it; if the customer chooses a package provided by a third party, he will be switched to a custom video source. The two possibilities are outlined below.

Standard video source

The standard video source offers two basic levels of control to the customer, by making use of the keys on the customer's SSN keypad. The PLAY, STOP, ⟨STEP⟩, ⟨SLOW⟩, ⟨⟨SEARCH⟩⟩ keys offer simple single-key control over a programme, as indicated. With certain types of programmes, slow motion and step frame are not possible.

When a customer presses the FACILITIES key, it causes a menu of options to be displayed to him, including: change current disc side (if programme occupies more than one disc side); random access to a required frame number (CAV discs) or timecode (CLV discs); random access to a required chapter (if disc has chapters on it); displaying or clearing an index of current frame number or timecode and chapter number (if available) on top of the video from player; changing audio status: either the left or right or both channels can be sent to the user; simple title-specific menus (optional).

The last option, title-specific menus, allows the user to get away from Laservision technology-dependent concepts such as frame number, timecode, disc side or chapter number and, instead, go to a point in the programme by choosing the content of that part from a simple list. For example, if the videodisc *British Garden Brids* were on the library, the title-specific menu would probably offer menus of particular birds, a tennis training programme would offer menus of particular strokes of the game to learn about, and so on.

Custom video source

The custom video source facility is provided to allow interactive programme packages to be offered on the library. These packages will mostly be provided by a third party, so once a session has started on a custom source, the user interface is defined by the programme creator. The package could be anything from a simple linear programme to a sophisticated interactive video learning package, driven by a local PC, or even coupled via a local micro to a remote computer. All possible one-byte keystrokes can be used (including output from a full alpha-numeric keyboard) except those codes corresponding to the FINISH and AUX keys which are filtered out before being passed to the video source; the FINISH key forces the session to be aborted.

The main difference between the standard video source and the custom video source, as far as the customer is concerned, is that the standard source has a fixed reaction to all keystrokes for all programmes used on it, whereas the custom source reaction will vary from package to package.

A number of custom video sources have been considered. A copy of the BBC Domesday system has been interfaced to the trial video library and is connected to the Westminster network. Some of the programmes from the Interactive Video In Schools (IVIS) project (a project administered and funded through the National Interactive Video Centre, London) have also been investigated for the video library, and it is intended to connect them to the Westminster system at a later date. In addition, it is intended to install a copy of an interactive video training programme developed by British Telecom for its own staff.

Technical overview

The philosophy behind the library design is to provide video sources on a one-to-one basis to customers during a session. The video sources have been dimensioned to meet expected traffic levels, which in turn were derived from a market analysis performed at the start of the design stage.

The multiplexed video and audio output from a video source is connected to a large three-level baseband video switch, whose outputs are connected to the outgoing video channels from the library. During a session, the switch connects the appropriate video source to the outgoing channel allocated to the customer for that session. The inputs to this switch include not only the video sources but also a set of text generators which are used for requesting from the customer, at the start of a library session, the identity of the programme he wishes to watch.

The library is controlled by a small 16-bit computer, which has a direct link to the main control system of the network. This computer communicates with the video sources, the logon text generators and the video switch via a control bus. The computer is responsible for a number of tasks, including setting up and clearing down sessions, keeping databases on programmes, copies and video sources, controlling the logon sequence via the text generators, allocating sources and programme copies to customers, passing on customer keystrokes to the appropriate video source, switching the required text generator and then video source to the given channel allocated to the customer, and deriving accounting information for onward transmission to the network administration computer. Figure 25.1 shows a schematic of the library.

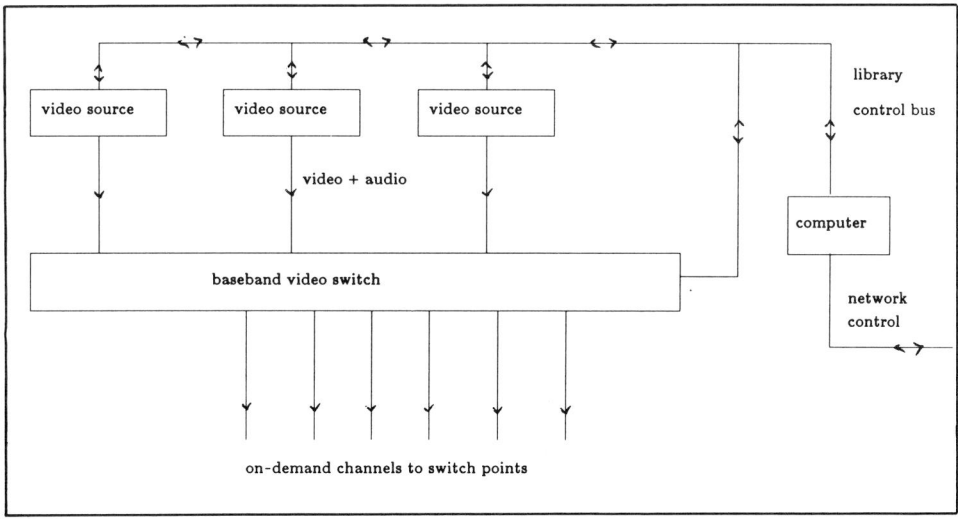

Basic library structure

Figure 25.1

The video source can be any device which delivers video and optionally accepts keystrokes from the customer: this design opens up the way not only for some form of standard source (as used already for simple Laservision videodiscs) but also for third party produced systems, as long as the packages can be controlled from the standard cable-TV keypad (or an optional alphanumeric keypad) and produce standard PAL-I video. Although that definition excludes, for the moment, touch-screen, mouse and bar-code reader systems, many packages now being developed could be run on the library with little modification. The hardware can use any type of video machine, from computer graphics through tape to VHD or Laservision videodisc.

The effective interactive video system set up by the library once a session has been set up, is shown in Figure 25.2

The standard Laservision video source, which is designed to cater for the complete consumer Laservision catalogue, offers the following: simple control by dedicated keys (stop, play, rapid, slow and step, forward and reverse); random access to any disc side, chapter, frame/timecode, using guidance displays control over audio channels and display of index; and the option of title-dependent menus, offering single-key access to named portions of the programme.

The standard video source can access up to four disc sides representing a complete programme, and provides textual guidance throughout the session. It comprises a teletext-grade caption generator, two small micros and a videodisc player, and implements the library-specific keystrokes as the Laservision discs allow: Constant Linear Velocity (CLV) discs, which are used mostly for films, do not allow Step frame or Slow motion, so when these keys are pressed by a user, a one-line caption, warning him of this, is overlayed for a few seconds on the video. The Stop function with CLV discs is implemented by pausing the player and displaying the caption Paused to the user.

The title-specific menus are held in an independent database, information from which is downloaded over a separate data communications link to the video source at the same time as the first disc side of a programme is being loaded into the player and being spun up to speed at the start of the session.

Automatic disc handling has been developed for the standard Laservision catalogue programmes, so that a large range of titles can be offered to customers at reasonable cost. The handlers come in modules of six players with access to 102 discs, with three such modules mounted in a rack of roughly 2 × 2 × 2 metres. The handler has been developed for BT by a British firm with long experience in the field of automatic disc handling.

Laservision videodiscs were chosen as the standard video media carrier as they offered the best

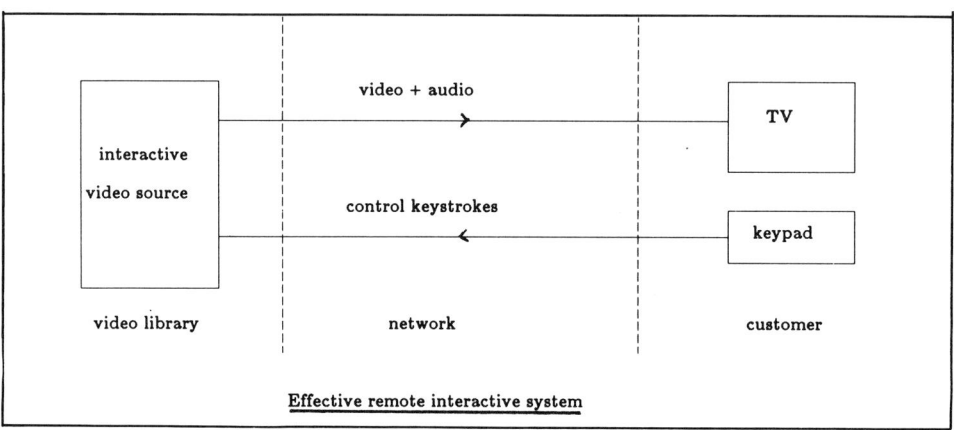

Figure 25.2

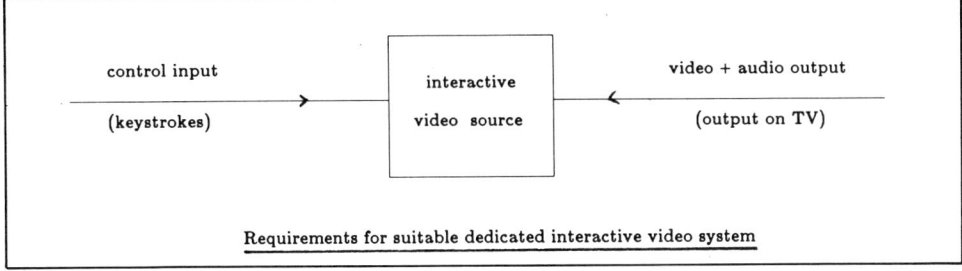

Figure 25.3

compromise in terms of cost of disc, cost of player, robustness, quality of signal, ease of handling mechanically and availability of suitable material. Until very recently, Laservision was really the only contender, the other videodisc formats (CED and VHD) having virtually or completely fallen by the wayside and having considerable limitations for the video library application. The new formats of optical disc, such as CD-I, CD-Video and CD-DVI open up new possibilities for the future, but at present have limitations for the existing system.[3] CD-I will only display graphics and still pictures, so standard films are not available on this format. CD-DVI is at present only available as a lab prototype, and although showing much promise, cannot be considered for the existing video library design in view of technical risk, availability and commercial cost. CD-Video, offering various sizes of analogue video with CD digital audio discs up to 30 cm in diameter, is becoming a contender, although the 30 cm discs are physically identical to the Laservision videodiscs and only differ in the coding of the audio channel. New CD-Video disc players would be required, along with modifications to the automatic disc handling equipment, to cope with the full range of CD-Video discs.

As well as accessing standard Laservision programmes, a customer can also gain access to other more advanced interactive video hardware, as mentioned above. The video library design allows for any custom interactive video hardware to be connected with few provisos; these are that the output signal must be able to be transmitted as PAL-I video, a suitable control interface be agreed between the hardware and the video library control bus, and the customer must be able to interact with the programme using either the simple system keypad, a full alphanumeric keyboard, or a compatible custom device provided for him (see Figure 25.3).

The sort of programmes envisaged in this category include: video encyclopaedia (visual database including maps, drawings, diagrams, manuals); remote viewing of famous art galleries; directed training packages (including salesman, maintenance engineers) partial replacement for evening classes for many subjects; replacement of, or an addition to, the broadcast scheduled educational programmes (such as the UK Open University broadcasts), allowing viewing in more social hours and self-paced study; DIY manuals on many subjects for the home.

Current status of Westminster network

The Westminster cable-TV system continues to grow as more of the franchise area is covered and more customers are connected. A large number of standard cable-TV channels are available to users, and in addition local videotex services are provided, perhaps with a direct link to Prestel.

A small-scale video library trial took place in April 1987, using Laservision films from the available catalogue. At the time of writing, a much more comprehensive commercial trial is planned for 1988/9.

Conclusions

The video library is perhaps the first implementation of an ideal way of offering video services, that ideal being to offer the services at the time the customer requires them. In addition, the video library offers the customer the capability of interacting with the video service, be that interaction merely to stop a film in order to answer a telephone call, or to react to part of a complex training or information package.

The popularity of video tape libraries testifies to the idea that people like to choose their own viewing, and watch these choices at times convenient to them, as long as the cost involved is perceived as being sufficiently low to warrant that convenience. If most cable services were able to be offered on-demand to all customers at a reasonable cost, then cable-TV would indeed offer significant advantages over broadcast and satellite services; if high-quality wide-ranging interactive information packages were available at reasonable cost, then cable-TV could also offer significant advantages over standard domestic videotape.

If all cable services were to be offered on-demand to customers on a one-to-one basis, that would indeed 'burn up the bandwidth', and the cost today of doing that is probably far from reasonable. The BT video library has been developed as a pioneer in this new area by offering some on-demand full-video interactive services in addition to the otherwise scheduled cable video services, and can be provided to customers on BT's switched-star network at a reasonable cost.

ACKNOWLEDGEMENT

Acknowledgement is made to the Director of *Research and Technology*, British Telecommunications, for permission to publish this paper.

26

The opportunities for Information Providers on broadband cable systems

Alan S M Robinson

The cable industry is presently in a state of change not unlike the change that was seen prior to the Hunt Report in 1983. This change has been brought about out of necessity for the survival of the industry and, one hopes, a new look at what the eventual requirements and end uses of cable systems will be in the UK reaching into the 21st century.

In order to present and discuss the case for opportunities for Information Providers on broadband cable television networks, one must first consider the evolution of the industry to date, present economic forces and the alternative routes ahead. Because of the utility nature of broadband telecommunications systems and the regulatory environment in which they exist, any analysis of the industry must look to the past, present and future governmental policies to distinguish the probable from the technically feasible.

The opportunities therefore, will be considered on the basis of the intended past development of cable, the actual development and the possible future development of cable systems in the UK.

Intended development

The serious consideration of the development of broadband cable systems in the UK probably started with the publication of the Information Technology Advisory Panel Report (ITAP) in the spring of 1982. This report opened the way for further consideration of the development of sophisticated information delivery systems and led to the commissioning of the Hunt Report which was published in October 1982.

Momentum gathered for the development of a hitherto government-controlled broadcasting and telephone industry and a White Paper was published in April 1983 leading the way to the announcement of 12 broadband pilot franchises.

The premise upon which this commercial operation was promulgated was the development of Integrated Services Digital Networks (ISDN) and the creation of highways of communication funded by private enterprise on the back of entertainment television. Various substantial Anglo-American joint ventures came about in 1983, such as Racal-Oak, Plessy-Scientific Atlanta, GEC-General Instrument, UEI-Times Fibre, each of which purported to have technology to meet the requirements of this new government doctrine. Underlying the commercial funding of this industry, was the belief that the government could stipulate a technology in order to create an industry which would match its own end desires. This development was premised upon the rapid installation of advanced networks. The development of United Kingdom switched technology which would drive the ISDN systems development in the UK would be a leader in the European market. The industry would, of its own accord, create new and hitherto unseen interactive services and all funding would be provided by private investment seeking the lucrative returns from cable television entertainment.

The UK was to lead Europe, through liberalised legislation, and in the technology and installation of these new broadband cable systems. Only limited encouragement would be needed for the creation of interactive services, which would develop of their own accord where advanced systems were mandatorily installed and readily available by government regulation. It was assumed that ISDN would develop rapidly for these new broadband systems and research and development would move away from the current twisted-pair telecoms technology to the new broadband fibre or coaxial delivery systems. It was presumed that the cost of optical fibre would drop rapidly with demand and replace both copper and coaxial cable.

All of the foregoing presumed that the commercial demand for entertainment services would be so considerable and immediate that the cost of advanced two-way systems would be an insignificant burden to the builder and operator of these systems. In retrospect, considerable parallels to the videotext industries' development in the late 1970s can be drawn from that of the cable industry. A basic fallacy, when blinded by the prospect of new technology, is that 'if you put a PC on everyone's desk then surely electronic mail will flourish and old-fashioned paper will disappear'. Little thought was given in this exercise to the providers and users of the interactive services, the cost in management time as well as financial terms, and the attention span of commercial enterprises in assessing and developing specific services for industry and the public at large.

The intrinsic fact is that it is human nature (of most non-technical users, whether commercial or general public) to continue with established systems rather than expend energy to learn a new and often daunting system.

No universal standards were present at any level, with regard to the speed of delivery of these systems (that is V24, V22, 1200/75 baud, 4800 baud, 9600 baud) nor universal consumer protocols, nor indeed the necessary software operating systems, and therefore no economies in scale for research and development and full production were available.

The actual event

In the summer of 1983 some 37 groups rushed to submit applications for the 12 pilot licences which were to be granted, and 11 were announced in November 1983. At the time of the announcement no actual definition of the required technology had been settled (only switched star for a 23 year licence); the question of which equipment was most appropriate became a moving target with the Department of Trade and Industry (DTI) and system capability gradually emerged as the real unspoken goal. It was, however, hard if not impossible for the DTI to move away from the statement that 23 year licences would be granted only where switched technology (whatever that might mean) was actually installed.

By early 1984, a number of the Anglo-US joint ventures began to look shaky as inevitable tensions between the US commercialism in production techniques *vs* the Anglo desire for theoretical perfection emerged. The first to go was the GEC-Jerrold joint venture followed, in the late summer of 1984, by Plessy-Scientific Atlanta joint venture and then, finally, the Racal and Oak Industries venture broke too, despite substantial orders of equipment from British Telecom. Only the UEI-Times Fibre link has survived and this is in a very modified form with a totally English piece of equipment.

The licencing of new systems was slow—bound up in the fight between British Telecom and the DTI over the terms of the 1984 Telecoms Act—and the loss of capital allowances was a severe blow to the industry. When construction started on broadband systems in 1985, it was with a much-changed industry seeking to pilot systems on limited equity and little or no debt support from the City itself.

The race was on in these systems to secure paying customers at minimal capital costs. Management and resources were too stretched in this fledgling industry to commit to the vagaries of interactive services and, as yet, the manufacturers of CATV equipment had failed to solve the basic problem of how to provide voice telephony on a broadband system without resorting to a twisted-pair overlay.

This is not to say that cable operators totally ignored the concept of interactive services, however. It became clear, at least in Croydon, that the operator would not only have to provide the highway, but also to create the service and encourage the customers. There were few suppliers queueing up to get on the system and, besides which, the physical reach of the system was very limited.

The voice requirement

The licence restrictions, caused by the BT-Mercury duopoly position, requires a broadband operator to have an agreement with either BT or Mercury before providing voice telephony services. This was thought to be of little consequence with regard to the provision of non-voice interactive services. However in practice this has not proven to be so and voice has become a necessary secondary requirement to many primary data services (for example the provision of automatic ticket barriers connected to a centralised monitoring station requires an emergency voice return circuit in case of failure of the machine). Only two cable operators have today reached agreement with Mercury for the provision of local voice telephony, and as yet no such service is being provided anywhere in the United Kingdom on a broadband system. The question of retained national accounts by Mercury in sites where more than 30 lines are present is of major concern, and is a fundamental stumbling block to the agreement of operators to the Mercury contract.

Croydon Cable Television was perhaps unique with the installation of both an A and B 450 MHz bi-directional trunk from the outset of the construction of the system, thus providing a totally separate and discreet institutional network. As early as 1985, the council was actively encouraged in the development of uses for this institutional network. CCTV funded an overseas trip for the Chief Librarian to the United States to discuss the use of institutional networks with such systems as the Manhattan Cable system and the Boston system. The overall result of this exercise was to identify the need within the Council for dedicated and experienced staff to plan the use of such a network and assess the cost savings which could be produced from such a relationship. No such liaison has resulted after a period of some two years. CCTV would have willingly passed this task to a third party, to develop in conjunction with the Council and to share in the risks and profits by development of the use of this institutional system. CCTV did in fact co-fund a management report to the Education Department which identified the potential market for high-speed, high-volume information distribution to the school administration offices and libraries, as well as distant-learning and software downloading to the pupil's home. A three-year trial was proposed in this report, which has been put on hold due to lack of supervisory management or the commitment of funds by the council.

This lack of development of the potential of cable systems is not just confined to independent or non-BT franchises. Only in Westminster (one of five working BT franchises) have two-way interactive services been piloted in the limited form of a video library, on-demand music and limited consumer shopping. Again, the provider of the service is the costly and limiting factor.

One-way semi-interactive information services (such as full-screen teletext and photovideotex advertising channels) have been more prolific and successful. The concept of classified advertisements purchased at a local newsagent was developed by DIVERSE and piloted with reported success in the Reading system, with an estimated 3,800 subscribers. Again, the physical coverage of the system must be a major consideration when assessing the possible negative effects of introducing a service prematurely.

The picture portrayed may not seem encouraging, but it is not altogether surprising at the birth of an entirely new industry and delivery system, and is far from hopeless. In fact, one might say the only way forward is upwards and one can identify the three primary requirements for the development of information services. These are greater physical coverage of the homogeneous area by the broadband cable system, relief from inappropriate regulatory restrictions, and users knowledgeable about the services that can be provided on broadband systems. In this respect, local authorities, education districts, transport systems and emergency services are ideal long-term users of cable information services.

The possible future developments

The government has commissioned PACTEL to review the broadband requirements for the next two decades and what delivery options are available to meet these needs. It is apparent that, if Britain is not to fall behind developments in Europe, the practical means of ensuring that broadband systems are physcially installed, must take place. Proposition 15 of the Peacock Report raised the spectre of a national fibre grid, perhaps installed by British Telecom. It is questionable whether this would be a positive step forward in a marketplace now driven by private funding. However, for the cable industry to progress and the support of the city be gained in funding an industry which is primarily of a utility nature, without the guarantee of revenue, five fundamental questions must now be addressed and answered by someone authoritative in government. In fact, these questions have needed to be answered since the abolition of capital allowances in 1984. They are:

> Is it still the present government's policy to get broadband cable into the ground as quickly as possible?

> What are the intended uses for this cable?

> Is video and telecommunications switching at the same physical location necessary and practicable in the light of ISDN developments in the past three years?

> Is it intended that the present broadband cable systems be realistically interconnectable?

> Will the DTI entertain an immediate re-examination of current broadband cable system design, as an absolute essential in the light of operating experience and the development of compatible transmission techniques, such as MMDS, M3VDS and cellular radio?

It is now essential that we address the end objective of our industry in order to chart its future. If it is only to deliver visual entertainment to private households, there is certainly no need for the costly and complicated systems mandated by government today. Perhaps even alternative delivery systems could be more effective with the present high cost of underground construction in the UK. However, if the end point in the year 2010 is very high capacity, interactive systems capable of delivering integrated, voice-image-data and text to one screen, then the progressive commercial development of systems and available technologies must be planned and implemented in stages.

Cable television needs coverage and customers of all types in order to progress to these future stages. The industry exists in an environment of growing competitive distribution means, and has yet to reach a point of critical mass in order to ensure its optimum potential. Therefore I have been at the centre of considerable pressure on government to allow the use of Multipoint Microwave Distribution Systems (MMDS) in broadband franchise areas.

MMDS could provide up to 12 video channels, as well as audio and very high speed data channels, to an entire franchise area within 120 days of its establishment. This would permit the rapid exposure of a limited cable service (entertainment and data) to the entire market-place.

The pending advertisement by the Cable Authority of the Birmingham franchise, an area of some 380,000 domestic homes, plus considerable commercial and industrial premises, provides a focal point for a review of a number of government policies and exemplifies the telecommunications opportunities potentially available to new broadband systems in the 1990s.

The construction of the Birmingham system comes at the time of the BT-Mercury duopoly review; should the government relieve the restrictions presently placed upon broadband operators, Birmingham could stand as a self-contained telecommunications area, with international uplink facilities and able to provide very high-speed secure transmissions across a unified area. The use of MMDS in the initial phases of this project is of high importance to the industry and to the securing of finance of this £120 million project.

In conclusion, the cable industry in the UK is going through a process of necessary technical re-evaluation, in order to place itself in optimum position for the emerging competing distribution means of the 1990s. The unification of technical standards and the emergence of Information Providers willing to act in the third party role, and utilise the distribution provided by cable, are paramount to the economic development of this industry, and information technology in the UK.

27

The world market for broadband applications

David Rumble

What applications?

There are many ways to describe or classify the applications and services that might require a broadband communications medium. The usual first step is to consider a few relatively obvious services: videotelephony, videoconferencing, TV distribution, CAD/CAM, LAN interconnection, and so on. A little further thought will add the requirement to consider the main sectors in which these applications may be used: home, business and education.

In its role as part of the co-ordination of the CEC's RACE programme PA has worked together with the PTTs and equipment manufacturers to develop a taxonomy of broadband applications and services which not only covers these dimensions, but also introduces a helpful distinction between the types of communication used. This also gives some indication of the sorts of topologies or network architectures that will be needed to support the services.

The prime dimension of this taxonomy describes the form of communication:

> Distributive (broadcast) services are those which only require the delivery or selection of a 'channel' from a variety available (perhaps with additional fee or subscription). Services could include broadcast TV of current or improved quality or High Definition TV for entertainment, education or for corporate video distribution, or audio channels. These could also be pay or subscription TV services. This class of services would also cover electronic newspapers and other electronic publishing forms such as teletex.

> Retrieval services are those which provide a selection from remotely stored material or information. Services could include film, video or audio programme libraries and document or information databases from broadband videotex to full text and image reference libraries.

> Surveillance services include remote oversight of vulnerable sites using full motion or slow scan video and other monitoring techniques.

> Messaging services include all movements of information or material which are not real-time or interactive, whether text, graphics, audio, still or moving pictures, as well as offline transfers of datafiles or software between computer systems.

> Dialogue services are those which require simultaneous bi-directional broadband communications. These include videotelephony and videoconferencing, and data services for such purposes as LAN interconnection, remote CAD/CAM, real time telemetry and remote control, on-line distributed databases, and more straightforward data applications that require either very rapid or high volume data transfers.

The service which is likely to be the largest user of wideband communications channels for the foreseeable future should not be forgotten: voice telephony. Recent PA studies have found that the market for high bit-rate transmission systems will be dominated by requirements for trunk telephony bearers well into the 1990s.

A particular point to note is that the first three classes only require broadband in one direction, the return (control) channel being potentially covered within telephony bandwidths: these classes of applications can therefore be supplied with simpler network architectures.

Which applications?

There is plenty of evidence of demand for video distribution services around the world, although the UK cable TV experience is not encouraging. Use of the other classes has been much more limited, essentially being confined to various pilot or trial systems around the world. A small number of commercial installations of interactive cable systems have been undertaken in the US, but none have looked like approaching profitability. There has been almost no use of videoconferencing in comparison with the potential market, and the existing uses are largely satellite-based.

The biggest disappointment to the Gee Whiz school of futurologists has been videotelephony (videophony), which has been imminent since I used to read *Boys Own Paper* articles on 'What life will be like in the 1970s'. There has yet to be convincing evidence of demand, and there is little perceived *need* for video in day-to-day business communications. As we will see later in the paper, even the fans of videophony (such as the Bundespost) only see a potential mass market if videophony costs are no more that two to four times telephony costs.

If there is no mass demand for videophony, the market for dialogue broadband services is left to data applications such as CAD and mainframe or LAN interconnection. However, mainframe sites are counted in the thousands in most countries and only tens of thousands across Europe or the USA. Sites currently using CAD are even lower numbers.

MAN/WAN/LAN interconnection may bring significant growth in the longer term—possibly with the 100 Mbits/s + protocols such as FDDI or IEEE 802.6, but more likely with 10 Mbits/s protocols such as the IEEE 802.4 base for MAP. Again this implies much lower demand for bandwidth than is potentially required for video distribution (although circuit occupancy for these applications is effectively continuous).

How?

This dominance of broadcast (and possibly retrieval) services has significant implications for viable network topologies in the medium term: most of the demand can be satisfied with relatively cheap tree-and-branch analogue networks, and most of the remainder by 'ministar' networks where control information is returned to the switch and suppliers via the telephone network.

The second major factor which will determine the structure of wideband networks is the tension between the provision of bandwidth and the development of signal compression techniques. Projects within the RACE programme have suggested that current TV quality can be achieved with transmission rates as low as 20 Mbits/s, and laboratories around the world are contending that they can squeeze acceptable videophony into 56 Kbits/s.

It seems likely that the cost of signal processing will continue to fall in price at a faster rate than transmission costs, not least because of the difficulty of significantly affecting the basic network installation technology (trench digging!). Widely differing positions are being taken up on the issue of compression, but the Bundespost is the only significant remaining advocate of the long-term viability of uncompressed services such as 140 Mbits/s videophony.

Similarly, it is unlikely that the current CAD file sizes will remain untouched by data reduction techniques if the exchange of CAD files becomes a very frequent requirement in manufacturing industry.

Taking these factors together, a widening gulf is apparent between the bandwidth and network architecture requirements for business and home markets for broadband which, left purely to commercial considerations, would militate strongly against a common integrated broadband infrastructure.

At present the market for business data services at 48 Kbits/s or more is miniscule. The Bundespost record only 400 connections in Germany, 0.2 per cent of data connections: the UK market is not significantly larger. In both countries the evidence is that the majority of even these low numbers of circuits are in fact used to multiplex lower speed connections. The spread of ISDNs will cater for much of this existing market and its growth. ISDN at 64 Kbits/s will also cope with electronic mail and electronic document interchange (EDI) with sub-second transfer times per document. The business broadband market outside the limited CAD area may well, therefore, be quite small for some time to come.

The market for broadband services to the home is potentially much larger, but the potential revenue from each household is low, and such services compete with many other services and products for a share of personal disposable income. It is a rare family that is prepared to devote more than, say, £50 per month to any one source of information or entertainment, and the potentially huge costs of provision of the basic broadband network infrastructure will eat up a large proportion of this sum, leaving very little to be shared out among the service providers.

Who?

Who, then, will pay for the provisions of broadband communications networks? In the face of the factors described above, it is unlikely that any entrepreneur or venture capitalist would find the market for an integrated network attractive. The provision of integrated broadband networks is likely to cost billions, with early potential revenues failing to match this by at least an order of magnitude. The UK experience suggests that launching new cable TV enterprises in a market with acceptable off-air picture quality and with a high penetration of VCRs as potential sources of home feature film viewing is at best a nerve-wrackingly risky business. Where cable TV networks already exist, it seems that the economics of their replacement will tend to lead to renewal as tree-and-branch or ministar systems. Any commercial provision of business broadband is likely to proceed on separate, lower-speed, high-quality networks.

The most optimistic forecasts for use of broadband services published this year come from the Deutsche Bundesministerium fur Post und Telekommunikation, in their 'Medium Term Programme for the Development of Technical Communication Systems'. This document quotes a variety of forecasts for subscribers to switched broadband services in Germany, ranging from 100,000 to 1.2 million in 1995, and from 500,000 to 5 million in 2005. The main factor differentiating for forecasts is the assumed cost to subscribers: in the lower forecasts the assumed cost to the subscriber for videophony service is four times current telephony costs, in the higher forecasts twice telephony costs.

The implication which may be drawn from the Bundesministerium's forecasts is that the price cross-elasticity between videophony and telephony is high, and that cost factors as low as seven to ten times telephony costs will reduce the market for videophony to the gadget-collectors, and effectively the forecasts also imply that the mass market price for a videophone must be well below £1,000.

If the entrepreneurs will not do it, the only chance for development of a coherent, integrated broadband network is by proactive, technology-led investment by Governments and PTTs on the grounds of the beneficial economic effects on industry of a better telecommunications infrastructure. This is the line followed by the Bundesministerium, and similar considerations have been explored by PA in the last few years in studies for the Belgian Science Policy and Planning Service, for the Zegveld Commission set up by the Netherlands Government, and latterly for the UK Department of Trade and Industry's Steering Group on the Evolution of the UK Communications Infrastructure (the Macdonald Committee).

Where? When?

The aim of the CEC's RACE program is to seek the establishment of a Europe-wide Integrated Broadband Communications Network (IBCN). At present that aim seems a long way off. The Bundespost will overlay a network on their ISDN; Belgium and the Netherlands will consider some investment in upgrading Tele-Distribution Networks as they are replaced; France will announce a plan to give away broadband terminals to replace the Directory Enquiries service. Elsewhere, the broadband market will be dominated by high-order telephony bearers and the use of satellites for TV distribution; data services in the 2 Mbits/s–10 Mbits/s speed range are likely to be mainly provided over separate dedicated private circuit networks almost everywhere.

For the rest of this century there will continue to be a variety of experimental networks from the Japanese INS to BIGFON and Biarritz, but the investment by the communications authorities will be dominated by the digitalisation of telephony and the narrowband ISDN. Timescales for network changeouts are long—the Bundespost aim is full digitalisation of telephone exchanges by 2020—and the renewal of local cable with broadband in fibre or other form is likely to be spread well into the middle of the next century.

In the longer term, the current experiments will undoubtedly lead to the technological advances necessary to provide readily a truly integrated communications network cable of dealing simultaneously with the demands of funds transfer and HDTV at a cost which makes it plausible for mass market use in competition with broadcast TV and other dedicated solutions for broadband applications. (In my own view that solution seems most likely to lie in packet or asynchronous time division (ATD) switching techniques—but that is not the subject of this paper.)

That technology should be settled enough in time to form the basis of the generation of switching systems that will replace System X, that is the network will mostly be installed in the early part of the next century. In the meantime the world market for broadband applications is likely to be restricted to experimental systems and a limited set of high-value business applications such as videoconferencing between widely separated corporate sites.

As noted earlier in this chapter, these applications will almost entirely be carried by utilising spare capacity on the bearers used for the telephony networks. In some major city centres, it will also be possible to utilise specialised fibre networks installed for business communications, such as BT's Flexible Access System and Dealerinterlink services in the City of London. Hopes for rapid development of integrated broadband services based on entertainment cable systems are fading even further as such systems come under competitive threat from microwave video distribution systems, direct broadcast satellites, and broadcast subscription television services.

The current (1987/8) reviews of policy in the UK will provide pointers to the likely direction of developments through the rest of the century. Any pullback from support of local cable franchising as the prime means of developing both free-enterprise television and local telecommunications competition will confirm the restriction of wideband communications to an adjunct of the telephone network.

28

Satellites and small dishes: business or pleasure

David Shorrock

Introduction

The role of satellites as communications media is still evolving. Traditionally used by the telecommunication administrations for shipping telephone traffic, data and television internationally, they are increasingly being employed by large corporate users of telecommunications as well as broadcasters and information providers. This expansion of use has occurred because of technological advances and overall cost reduction in satellite communications, as well as the removal of many of the regulatory restrictions.

The major expansion of activity has occurred in the USA, particularly in the telecommunications field. The deregulation of the telecommunications industry in 1983 allowed new freedoms and introduced new carriers, some of whom based their services on satellites. These satellite carriers offered services ranging from reselling transponder capacity on satellites which they flew and operated, through to the provision of the transmitting and receiving earth stations. These freedoms also extend to individual users who, subject to (virtually automatic) FCC approval, are allowed to install and operate their own transmit-and-receive earth stations and can therefore construct their own private telecommunications networks, independent of the terrestrial carriers. These users typically claim network cost savings of between 15 to 30 per cent against their terrestrial counterparts.

But satellites have not been confined to business, they have become a consumer product. Over the last five years, Americans have installed tens of thousands of 'backyard dishes', to listen in to satellite transmissions, and particularly those channels carrying recently released films, that are being distributed nationally by satellite, for redistribution by local cable TV companies. The freedoms enjoyed in the US are in stark contrast to Europe, where satellites are still seen as a government or PTT monopoly.

Regulation of satellites in Europe

In Europe, the regulation of satellite services is dependent upon a number of factors, the first being whether the service to be provided is national or international. In the case of national services, these are usually under the control of the government or a government appointed agency. International services are permitted by bilateral agreement between the countries concerned. However, a further complication is the difficulty in regulating services aimed at reception in one country, but which can be received in other countries, possibly not party to an agreement.

A further division is made between 'broadcasting' and 'fixed transmission' satellites, which often come under different departments of a national government. It is common for the fixed satellite services to be classed as telecommunications and they are, therefore, the sole responsibility of the PTT, whereas broadcasting services, which are usually classed as entertainment, are under the auspices of either a government-controlled or an independent broadcasting agency.

In general therefore, the provision of satellite links for telecommunication services is the exclusive privilege of the PTTs within the various countries in Europe. As Figure 28.1 shows, in most countries this applies to the provision of ground equipment for both transmission (the uplinks) and reception (the downlinks). However, there has been limited relaxation in some countries for downlinks only.

The distinction between broadcasting and fixed transmission satellite services becomes blurred when the new small dish services, very small aperture terminals (VSATs) or microterminals are considered. Then it is often the application of the service—entertainment television, news services, financial information, corporate data, etc—which somewhat arbitrarily decides which department has responsibility for the service.

Country	Downlink only	Uplink & Downlink	Comments
Belgium	PTT	PTT	-
Denmark	PTT	PTT	-
France	PTT or Private	PTT	Licences may be issued for private uplinks
Germany (FRG)	PTT	PTT	-
Greece	PTT	PTT	Position unclear but no formal policy to license private operation
Ireland	PTT	PTT	Private satellite operations under consideration
Italy	PTT	PTT	Private operation of downlinks may be permitted in certain circumstances
Luxembourg	PTT or Private	PTT or Private	Private operation permitted in limited cases
Netherlands	PTT	PTT	Private operation of downlinks may be permitted in certain circumstances
Portugal	PTT	PTT	Position unclear but no formal policy to license private operation
Spain	PTT	PTT	-
United Kingdom	PTT (BT or MCL)	PTT (BT or MCL)	6 private uplinks licensed, downlinks permitted

Responsibility in Europe for the provision of satellite services

Figure 28.1

What is a VSAT?

The last six years have seen the development of large scale, small dish satellite data broadcast networks: there are currently in excess of 35,000 small dish (60 cm diameter) receive-only satellite terminals installed across the US. These are used for distributing such information as stock market prices, news and corporate communications. These small dishes, referred to as 'microterminals', operate in the microwave C-band (4–6 GHz) and typically use a proprietary spread-spectrum technique. They are capable of receiving data at rates ranging from 300 bps to 19.2 Kpbs.

In the last three years in particular, microterminals have been joined by two-way, or interactive terminals, which are capable of both transmitting and receiving. In the US, these terminals are generally referred to as very small aperture terminals, VSATs. In Europe, the terms VSAT and microterminal are used interchangeably.

A typical VSAT network is shown in Figure 28.2 and can be seen to comprise a central hub earth station which transmits to and receives from a large number of remotely situated VSATs. The data network shown has a single host computer, but generally there is no restriction on the number of hosts that can connect to the hub and thus a number of applications and closed user groups can be served by the single hub earth station.

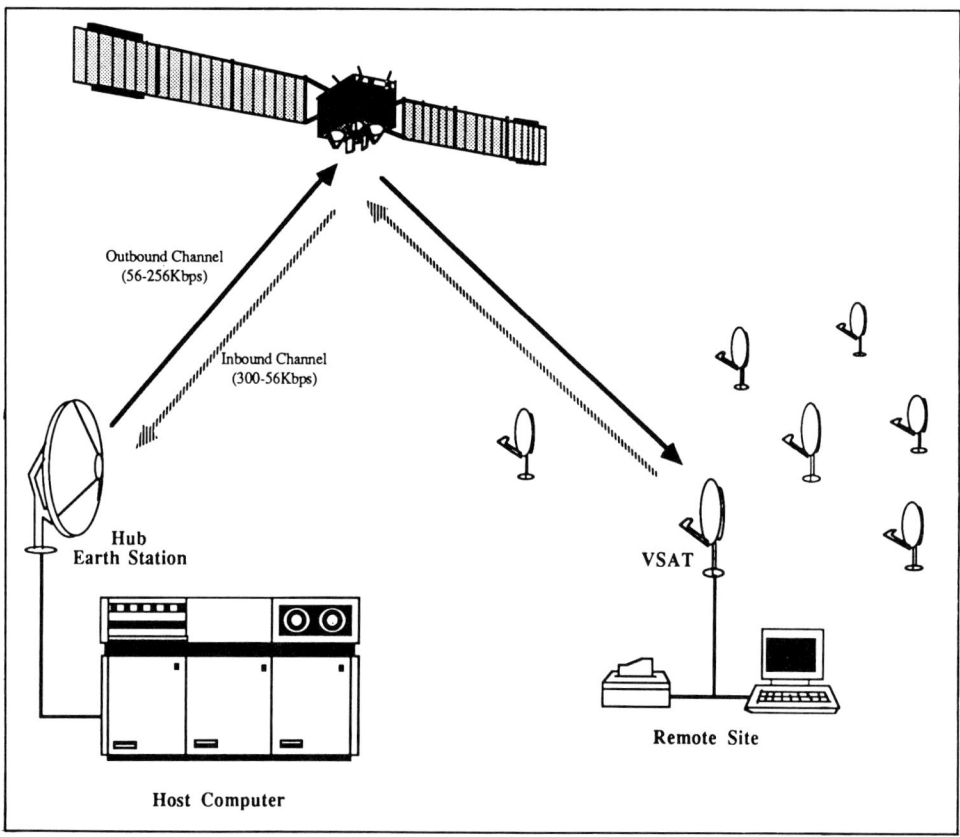

VSAT Satellite Data Network

Figure 28.2

The VSATs are typically 1.2 to 1.8 metres in diameter and are capable of receiving data at rates ranging from 56–256 Kbps, as well as having a transmitting capability at data rates varying from 300 bps to 56 Kbps. The exact data rates on the transmit and receive paths vary according to the vendors' equipment, although a feature of all VSAT systems is that the received data rate, the 'outbound' path from the hub earth station to the VSAT, tends to be higher than the transmitted data rate, the 'inbound' path. The reason for this asymmetry in data rates is that for small antennae (which by their nature have a wide beamwidth), it is necessary to restrict the transmitted power/bandwidth (and hence data rate) to avoid interference with adjacent satellites. Further, both the received and transmit data rates are rarely available at the full bandwidth continuously to an individual VSAT, but are shared between VSATs by means of an access scheme, such as demand assigned or time division multiple access.

The majority of VSATs offered by vendors operate at the higher microwave frequency band referred to as Ku-band (11–14 GHz). The major reason for using Ku-band for VSATs is to overcome the radio interference problems associated with the lower frequency C-band which is shared with terrestrial point-to-point microwave links and hence requires careful frequency coordination when siting terminals. It is for this reason that the Ku-band has been adopted as the primary frequency band for fixed telecommunication satellite services in Europe. Consequently, in Europe, both microterminals (receive-only) and VSATs will operate in the Ku-band.

There are approximately 25 VSAT vendors worldwide, the majority of whom are based in the US and Japan. There is no compatibility between products from different vendors, vendors using proprietary hardware and software. As yet, there are no internationally recognised standards for VSATs.

In addition to a data transmission and reception capability which may support compressed voice, many VSATs can be configured to include a video demodulator, so that they can receive analogue television signals, providing the television signals originate from the same satellite as the data.

The benefits of VSAT networks over terrestrial services are claimed to be:[1]

end to end service can be provided by a single vendor and since VSATs are located on a customer's premises, there is no 'last mile' to contend with,

annual costs of 25 per cent or more below equivalent terrestrial service costs,

VSAT based data services can offer high quality and availability (bit error rates better than 10^{-7}, availability better than 99.5 per cent),

due to speed and ease of installation (typically one day) and computerised configuration management, the network manager can quickly respond to changes in demand and service allowing VSAT locations to be moved or added quickly to the network,

VSATs can provide a variety of services from data to video, voice and facsimile, in an integrated manner. The mix of services provided can be changed as a customer's business needs change,

Technologies such as time division multiple access, TDMA, allow for the efficient utilisation of satellite capacity, so that a customer pays only for the capacity actually used.

Applications

The following examples of VSAT users, both information providers and corporate users, illustrate some of the benefits of satellites over terrestrial services.

Reuters: real-time financial information

Across the continental US, Reuters are serving approximately 5,000 customers with receive-only microterminals, situated on the customers premises. The services offered are the 'Reuter Monitor Services', real-time information such as general news bulletins, futures and options quotations, currency markets, stock market indices, OPEC and other energy related statistics, government reports, weather information. All the data is transmitted from the company's own uplink at the Long Island Technical Data Centre, via leased satellite capacity, to the customer's 60 cm diameter dish. The information is then displayed on a colour video monitor and personal computer with a 500 page display. The system is capable of transmitting data at up to 19.2 Kbps, equivalent to 20,000 words per minute.

The major benefits claimed by Reuters is that satellite technology provides the widest possible coverage for disseminating information. It is for this reason that satellites are favoured by broadcasters. In addition, unlike terrestrial networks, the number of receiving points can increase without affecting the network loading.

K mart Corporation: data and video for retailing

K mart is the seventh largest employer in the USA and has some 2,050 retail outlets. It is estimated that half the families in the US shop at least once a month in the chain. As part of $500 million five-year retail automation programme, K mart is introducing a VSAT network to link each of its stores with the corporate headquarters in Michigan.

The VSAT network, which is being supplied by subsidiaries of GTE Corporation, GTE Spacenet and GTE Telenet respectively, is estimated to cost $50 million, of which between $2 million and $3 million is for the hub earth station at corporate headquarters, and between $15,000 to $20,000 for the VSATs situated at each store.

The K mart network has been designed to support both two-way data and video transmission. Data applications include on-line credit authorisation and access to inventory data, either from individual stores or throughout the chain. The video facility offers full motion video broadcasts, both for presentation of merchandise and for corporate communications.

According to a K mart spokesman, the satellite network is the most cost effective method of providing communications to their stores: "The amount we pay for the satellite is virtually independent of the number of stores. The biggest expense is putting the dish on the buildings. But once you put the dishes up, the cost of data communications between the store and headquarters is negligible".

Other benefits claimed include credit card authorisation times reduced by between 80 and 90 per cent and improved communications between senior management in corporate headquarters and personnel in the stores.

Days Inn: data and video for hoteliers

Days Inns of America is a hotel chain of some 450 hotels comprising 63,000 rooms across 43 states and Canada. As a means of improving the reservation system, which was previously based on a central reservation bureau with dial-up to the hotels, Days Inn are installing a VSAT network to provide two-way data and one-way video. The system is estimated to cost $9 million and is being provided by GTE Spacenet in conjunction with Tridom Corporation, the manufacturers of the VSATs. Each hotel will have a 1.8 metre diameter dish, at an estimated cost of $7,500 per site.

The central reservation system, that sits on the host computer at Days Inn headquarters in Atlanta, is connected via leased terrestrial circuits to the shared hub earth station operated by GTE Spacenet in New York. Days Inn estimate that by using the satellite system, reservations times will be reduced from the current 40 to 50 seconds to under 10 seconds, and with projected cost savings of $300,000 per year.

Although the primary justification for the satellite network was the improvement in the reservation system, the network also offers the capability for providing in-room entertainment in the form of video transmissions via the satellite. Days Inn are currently negotiating with potential entertainment suppliers.

An additional benefit identified was the speed in installing the VSATs, compared with the typically three-month delay in obtaining telephone lines. With Days Inn looking to double the size of their chain, to 900 properties by 1990, this becomes an important consideration in integrating new hotels into the existing network.

Services in Europe

Given the regulatory restrictions surrounding satellite communications in Europe, there are currently few examples of users or information providers using VSATs. One, however (Polycom, described by Lionel Fleury in the following chapter) is an example that overcame the restrictions by forming a joint venture company with the French PTT. Given these restrictions, the European Space Agency (ESA) initiative to promote new satellite services on the experimental Olympus satellite looks particularly ambitious.

Olympus: promoting new satellite services

Olympus is an experimental high power communications satellite scheduled for launch in January 1989 and developed at a cost of £400 million by ESA. Olympus will act as a high technology demonstrator for potential new services and applications of satellite technology for the 1990s. For the first two years of operation, Olympus will be available to users free of charge to conduct experiments, although the users will have to fund the costs of the ground equipment themselves. At the end of the initial two-year period, users will be able to continue using Olympus as pilot services, on payment of a nominal fee.

Approximately 120 potential users have expressed interest in Olympus, the majority looking to conduct experiments on the direct broadcast satellite (DBS) transponder. Of these, the majority are educational and training institutions looking to conduct distance learning experiments through a medium which otherwise would have been prohibitively expensive.

One of the most dramatic uses of Olympus is that proposed by Birkbeck College, London, and Guy's Hospital. The project is aiming to compare and transfer psychiatric and psychology expertise in the control of intravenous drug addiction and its relation to the spread of AIDS. Using a transportable earth station, broadcast videoconferences will be held for two hours each week using video tapes of patients, superimposed text and graphics relating to test results, as well as four audio channels for audioconferencing in different languages. European participants in the project will include the University of Paris, the University of Geneva and the Max Planck Institute of Psychiatry, Munich. Doctors at these institutions will be able to telephone questions back to the studio during the broadcast and participate in the discussions, while watching the broadcast. The transmissions will be encrypted to maintain the medical confidentiality of the participants.

The project is being developed jointly by Dr Angela Summerfield, Senior Lecturer in Psychology at Birkbeck College, Dr Maurice Lipsedge, Consultant Psychiatrist and Professor James Watson, Chairman of the Division of Psychiatry, both of the United Medical and Dental Schools of Guy's and St Thomas's Hospitals. The importance of the project, according to Dr Lipsedge is that:

"methods and techniques in AIDS prevention are currently centring on the control of drug abuse as a major weapon to stem the progress of AIDS into the heterosexual population. Progress on methods is so fast that there is a need to assemble quickly on a European basis to transfer expertise and to tackle problems in a concerted European initiative.

European countries have for example experienced differing success rates in dispensing needles to intravenous drug users to help restrict the progress of AIDS. The satellite links will enable specialists all over Europe to debate and analyse the successes and failures of such policies and work towards a common way forward."

Dr Summerfield, initiator of the project, describes it as a continuation of a "policy of outreach into the community which through new satellite technology will cross national boundaries for the first time".

Conclusions

Satellites offer a potentially cost effective means of communications, particularly for information providers who are broadcasting to a large number of users, possibly over a wide geographical area. In the US numerous examples exist of such services, which either would not be feasible or would be prohibitive in cost if carried over terrestrial circuits.

Yet in Europe, although the technology is readily available, the regulatory regimes are restrictive. The current regulatory environment governing European telecommunications are denying both information providers and users a powerful and cost effective delivery media. They must change, or the European information industry will face even greater competition from its American and Japanese counterparts.

The European scene is not entirely bleak. As discussed by Marcus Bicknell in chapter 31, by late 1988 or early 1989 the privately financed satellite, Astra, will be providing 16 channels of television to a pan-European audience. This will be a powerful competitor to the various national direct broadcast satellites which can offer only three or four channels, championed by Jon Chaplin in chapter 32.

Perhaps the most encouraging development is the Olympus project and the opportunity that will offer for the development of new satellite services and applications. Innovative thought is being applied to the potential that satellites offer, as exemplified by the Birkbeck/Guy's videoconferencing project. It is, however, ironic that it is only the developed countries that can afford such projects, in contrast to the poorer nations of the world where the need is greatest.

29

Data broadcasting by satellite on Europe

Lionel Fleury

Introduction

Polycom, incorporated in January 1986, is a joint venture between Agence France-Presse and France Cable et Radio, the latter being a subsidiary company of the French PTT responsible for marketing the services of the French Telecom 1 satellite.

Polycom acts as the transmitting agency, in that it offers the data broadcasting service via the European Communications Satellite, (ECS) or Telecom 1 satellites, under a licence granted by the PTT. Information providers connect to Polycom's headquarters in Paris and, from there, data is uplinked to either of the satellites at data rates varying from 50 bps to 9600bps. Polycom intends to offer higher data rates from 19.2 Kbps to 64 Kbps, subject to demand. The claimed bit error rate for the service is 10^{-7} for 99.9 per cent of the time.

Polycom does not supply the 90 cm diameter receive-only terminals, but rather makes the specification freely available to manufacturers, from whom the users can then buy the terminals. Currently, only Matra has terminals in production, at a unit cost of approximately FF 50,000.

Commercial service commenced on the 15th February 1987 with the first customer, Agence France-Presse, using the system to send both text and newspaper-quality monochrome photographs to their subscribers in and across France. Agence France-Presse have installed 120 terminals in France, as well as individual terminals in Stockholm, Madrid, Barcelona, Belgium, Switzerland and Corsica.

A second Polycom user, broadcasting share price information, is the Paris Bourse. The Bourse service commenced in November 1987 and terminals are being installed at the rate of 10 per month. Polycom are extending the service overseas via Intelsat: to the USA in the first quarter of 1988 and to Africa, in the first quarter 1989.

Polycom development

Data broadcasting by satellite was identified as a market opportunity in the preliminary market surveys for Telecom I in 1979. However, for various reasons, the European satellite systems, including Telecom I, had not been designed with such services. These reasons included: the use of Ku band frequencies which required components that were more expensive than those for the more commonly used C band; the satellite access standard was defined as 64 Kbps single channel per carrier (SCPC), or time division multiple access (TDMA), rather than simpler, lower data rate access schemes; regulatory restrictions in Europe applied to the use of receive-only terminals.

Subsequently, many of these problems were overcome: the restrictions on receive-only terminals were relaxed in many of the European countries due to the pressures from the development of satellite television; development of similar services in the US; the development of competitive marketing of news and financial services in Europe; the European PTTs gave priority to high bit rate services, resulting in potential users of slow bit rate services being ignored; Eutelsat, the European satellite operating organisation, and the PTT agreed to consider offering non-standard services for satellite transmission.

These developments encouraged the initial shareholders of Polycom, Agence France-Presse and France Cables et Radio, to apply to the French PTT for a licence to establish, operate and market a data broadcasting service.

The schedule over which the Polycom services were developed was as follows:

January 1986: transmitting system and prototype receiving earth stations ordered.

February 1987: opening of the commercial service on the Eutelsat I-F2 satellite.

November 1987: opening of commercial service on Telecom 1.

January 1988: opening of operation centres in Washington and Frankfurt.

March 1988: service data rate of 64 Kbps introduced in addition to 19.2 Kbps service.

From the beginning of the project, one of the objectives of Polycom's founders was to establish a European system using the Eutelsat system of satellites and equipment manufactured by European suppliers, instead of imported equipment. This objective has had several consequences: it led to the definition and subsequent development of a system specifically suited to the European environment, with the technical support of the French PTT and Eutelsat; the receiving equipment cost was of a

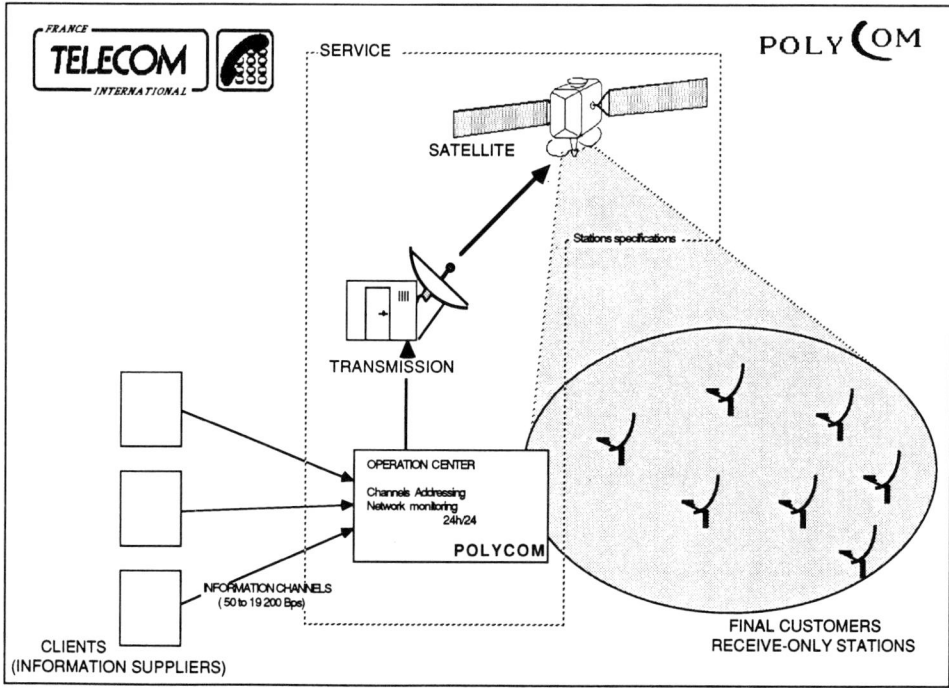

Figure 29.1

NEW MEDIA: COMMUNICATIONS TECHNOLOGIES FOR THE 1990S

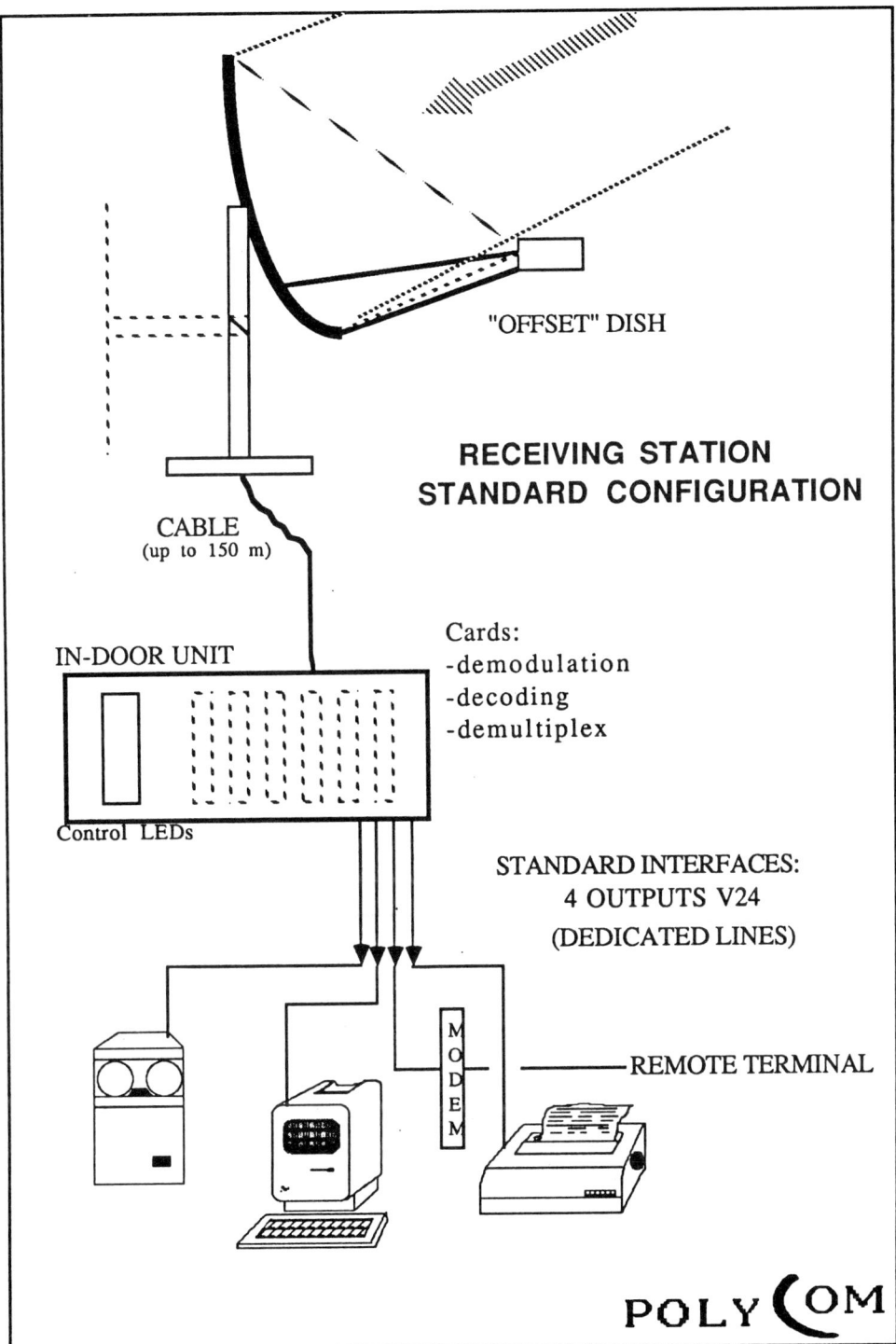

"OFFSET" DISH

**RECEIVING STATION
STANDARD CONFIGURATION**

CABLE
(up to 150 m)

IN-DOOR UNIT

Cards:
-demodulation
-decoding
-demultiplex

Control LEDs

STANDARD INTERFACES:
4 OUTPUTS V24
(DEDICATED LINES)

MODEM

REMOTE TERMINAL

POLY COM

Figure 29.2

higher initial price than the imported equivalent, although this has now been reduced by half and will fall further as greater quantities are produced. In addition, a policy decision was made to make the specification of the receiving equipment freely available to any customer or PTT administration, allowing them to obtain the equipment from any supplier. This policy can now be seen to have been in accord, if not a year in advance, of the Commission of the European Community's Green Paper on the liberalisation of telecommunications equipment and services.

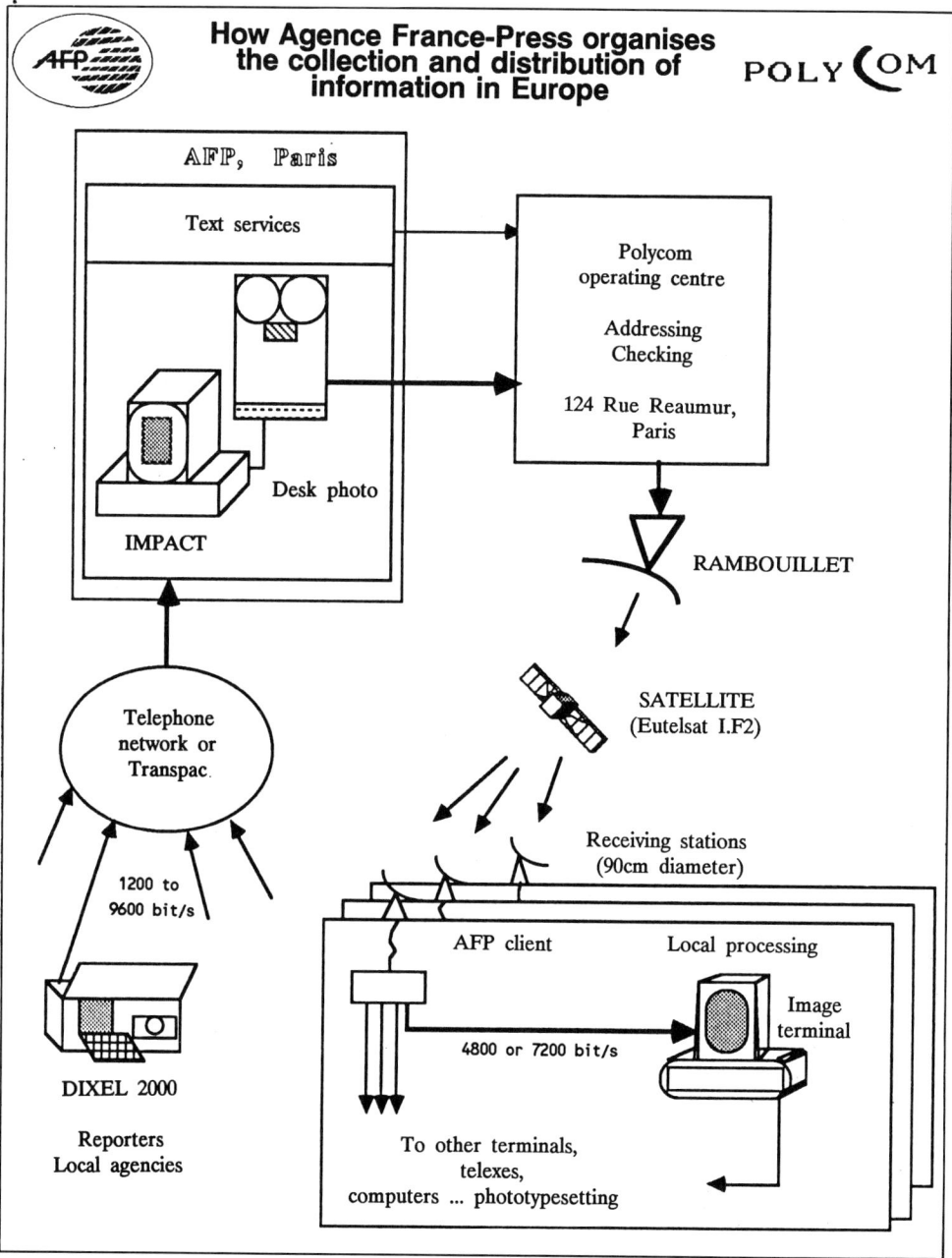

Figure 29.3

NEW MEDIA: COMMUNICATIONS TECHNOLOGIES FOR THE 1990S

The current service offered by Polycom corresponds with the needs of several European clients. However, following a policy of continuing improvements in the services offered, Polycom is pursuing the following objectives: to reduce the space segment and uplink costs; to reduce the costs of the receiving equipment and to develop simplified terminals for specific applications; to introduce a 64 Kbps service (and higher data rates) for specific applications such as remote printing; to diversify operation centres and serve other regions, both within and outside Europe.

Service description

The Polycom one-way data broadcast system is shown in Figure 29.1. Each client (information supplier) transmits information over terrestrial circuits to the Polycom operation centre, where it is processed for addressing purposes, multiplexed with other data from other clients and then sent over digital circuits to a PTT-operated uplink at Rambouillet, 50 Km outside Paris, from where it is transmitted up to the satellite. The information channels data rates are selectable from 50 bps up to 19.2Kbps, with extension to 64 Kbps. The entire system is operated and monitored 24 hours a day, every day of the year, from the Polycom operation centre.

The structure of a typical receiving station is shown in Figure 29.2. Each station comprises a 90 cm offset aerial, connected via cable to the in-door unit which performs demodulation, decoding and demultiplexing to reconstitute up to four information channels per receiver.

Figure 29.3 shows a typical user application of the Polycom service. Agence France-Presse (AFP), operate a photo service using a 7,200 bps channel on the system. Pictures are collected through the terrestrial networks, processed in AFPs central office in Paris, and then broadcast to subscribers through Polycom. The pictures are transmitted simultaneously with the accompanying text. The success of this application is witnessed by the fact that virtually every French newspaper office is equipped with this system.

Future interactive services

Interactive, or two-way, services employing VSATs are seen as a natural extension of Polycom one-way services. However, in Europe such services are problematic when compared with the situation in the US; the freedom to operate on-premises transmitting stations is restricted to the PTTs, the space segment cost in Europe is almost twice that in the US, and the European PTTs have developed extensive packet switched networks that offer competitive services to VSATs.

Polycom is continually reviewing the issue of VSAT services and would be interested in a joint commercial and technical relationship with a potential equipment supplier, if the regulatory and economic issues change. In the interim however, Polycom is examining composite solutions, that combine the broadcast capability of the existing Polycom service, with a return path from the microterminal to the host, provided by the packet network.

Conclusion

Polycom's goal is to provide the best service at the lowest price to its clients; it is also aiming to promote European technology and industry. These goals could be achieved more rapidly if the costs of the satellite space segment could be reduced further. Such an evolution would also guarantee the future of the European space industry as well as the European information industry.

30
Towards the intelligent satellite
Barry G Evans

Where are we now?

Satellite communications began in 1945 with the prediction by the now-famous science fiction author Arthur C Clarke that three satellites in geostationary orbit—that is, at a distance of 36,000 km above the equator where to an observer on the earth's surface they remain approximately stationary, Figure 30.1—could provide world-wide telecommunications coverage. This was no crystal-ball gazer's prophecy: Clarke had realised that a rocket, only several times larger than the V2s he had experienced with such horror in the Second World War, could in fact launch such a satellite. This, coupled with the

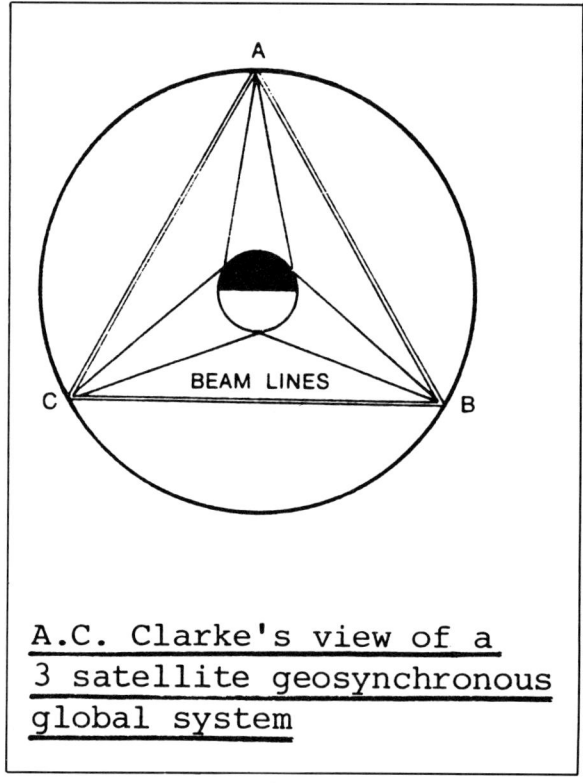

Figure 30.1

realisation that the innovations in microwave technology needed for the installation of radar could also be used to produce the communications equipment necessary for transmission to and from such satellites, made Clarke's prophecy no more than an educated forecast in telecommunications. However, Clarke produced the concept of the 'sky-hook' on which to hang our satellites. Since the early experimentation and trials via metallised meteorological balloons (the 'Echo' programme) and subsynchronous satellites (the subsynchronous era) we have passed through, or are about to enter five eras of satellite communications, as illustrated in Figure 30.2.

This paper will consider the driving force behind this rapid evolutionary path, which has been society's insatiable demand for more and more communications. More and more telephone (voice) circuits and increased video (broadcast) circuits have been the classical traffic patterns. Nowadays, these have been joined by the demand for more and more data services of all kinds: computer-to-computer services, electronic mail and electronic funds transfer, plus a range of new video services such as facsimile, videoconferencing, slow-scan television and many others, too numerous to mention. It must be pointed out that, besides the civil communications requirements, there have been other driving forces in the fields of military communications, especially in surveillance and in earth-resource sensing. This paper will be restricted to the civil communications field as, perhaps unlike other areas of technological advance, this has dominated satellite communications.

The problem faced by the satellite communications engineer has been how to provide for such traffic and services in an efficient and economic manner with one major constraint, that of restricted bandwidth. In order to avert chaos, and allow for international communications, the frequency spectrum has been planned by the International Telecommunication Union (ITU) and various portions allocated largely to separate communication systems or means of communication. An example of the allocation to satellite communications in the two bands currently used in civil communications (4/6 GHz and 11–12/14 GHz) is shown in Figure 30.3. It will be seen that, even with recent additions, there is precious little bandwidth into which to cram our communications requirements. So it is this, more than anything else, that has called for technological innovation and has presented satellite engineers with their challenge.

It took 23 years before Clarke's prophecy became a reality in 1968, when there Intelsat III satellites became successfully established above the major oceans. This had been preceded by earlier and smaller Intelsat satellites starting in 1964 with Intelsat I, and has been followed by larger and larger satellites culminating with the Intelsat V satellites carrying today's global communications and the giant (2 ton)

Satellite communications eras	
1. Subsynchronous era	1958–1963
2. Global-synchronous era	1964–1972
3. Domestic and regional era	1973–1981
4. Small station application era	1982–1990
(i) broadcasting	
(ii) business	
(iii) mobile	
5. Intelligent satellite era	1990

Figure 30.2

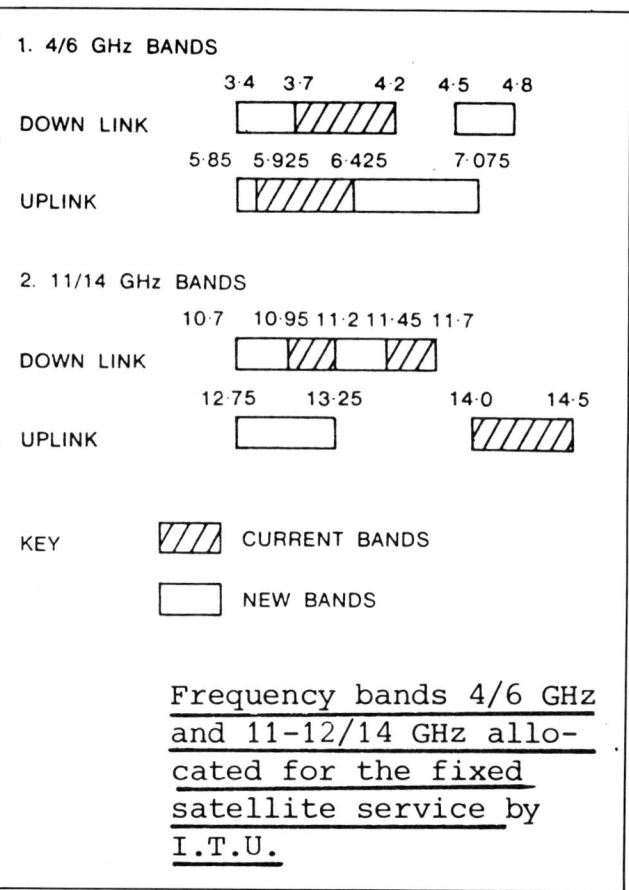

Figure 30.3

Intelsat VI planned for 1988–89. Figure 30.4 summarises the developments in the Intelsat global system and illustrates some of the innovations taken to meet traffic demands which were until recently averaging 25 per cent per annum. Returning to the eras of satellite communications, Figure 30.2 shows that, around the mid-1970s, the cost of satellite communications had reduced to a level which enabled regional and domestic systems to be established ecnomically. Whereas the large global system had developed around large 30 metre international gateway earth-stations, the domestic systems used smaller and cheaper 11 to 13 metre stations to serve the telecommunications needs of nations or collections of countries. This was done by either leasing transponder capacity from Intelsat or by launching a purpose built satellite (for example Indonesia, Canada, Arab League, Australia). The global and domestic-and-regional eras were the focus of the satellite business until the early 1980s. The satellites have developed to include several new techniques. Firstly, they were bigger and hence more powerful, which trades-off to more traffic capacity. Secondly, they had developed complicated antenna structures which allowed the valuable power to be constrained into the areas in which it was needed (shaped beams). Thirdly, frequencies were re-used by adopting orthogonal polarisations which also enabled more traffic to be carried in the resultant expansion of transponder. The latter two techniques are demonstrated in Figure 30.5 which shows the shaped beam and frequency re-use patterns of Intelsat V.

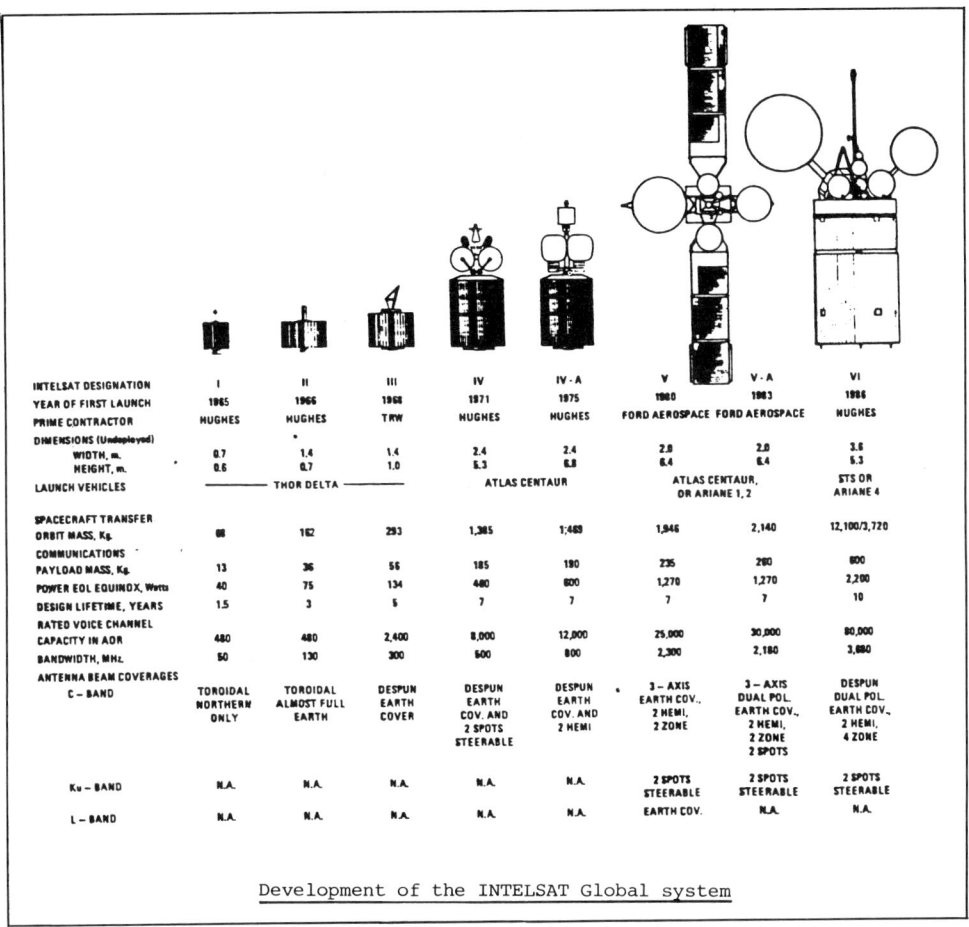

INTELSAT DESIGNATION	I	II	III	IV	IV - A	V	V - A	VI
YEAR OF FIRST LAUNCH	1965	1966	1968	1971	1975	1980	1983	1986
PRIME CONTRACTOR	HUGHES	HUGHES	TRW	HUGHES	HUGHES	FORD AEROSPACE	FORD AEROSPACE	HUGHES
DIMENSIONS (Undeployed)								
WIDTH, m.	0.7	1.4	1.4	2.4	2.4	2.0	2.0	3.6
HEIGHT, m.	0.6	0.7	1.0	5.3	6.8	6.4	6.4	6.3
LAUNCH VEHICLES		THOR DELTA		ATLAS CENTAUR		ATLAS CENTAUR, OR ARIANE 1, 2		STS OR ARIANE 4
SPACECRAFT TRANSFER ORBIT MASS, Kg.	68	162	293	1,385	1,469	1,946	2,140	12,100/3,720
COMMUNICATIONS PAYLOAD MASS, Kg.	13	36	56	185	190	235	280	800
POWER EOL EQUINOX, Watts	40	75	134	480	800	1,270	1,270	2,200
DESIGN LIFETIME, YEARS	1.5	3	5	7	7	7	7	10
RATED VOICE CHANNEL CAPACITY IN AOR	480	480	2,400	8,000	12,000	25,000	30,000	80,000
BANDWIDTH, MHz.	50	130	300	500	800	2,300	2,180	3,680
ANTENNA BEAM COVERAGES C – BAND	TOROIDAL NORTHERN ONLY	TOROIDAL ALMOST FULL EARTH	DESPUN EARTH COVER	DESPUN EARTH COV. AND 2 SPOTS STEERABLE	DESPUN EARTH COV. AND 2 HEMI	3 – AXIS EARTH COV., 2 HEMI, 2 ZONE	3 – AXIS DUAL POL. EARTH COV., 2 HEMI, 2 ZONE 2 SPOTS	DESPUN DUAL POL. EARTH COV., 2 HEMI, 4 ZONE
Ku – BAND	N.A.	N.A.	N.A.	N.A.	N.A.	2 SPOTS STEERABLE	2 SPOTS STEERABLE	2 SPOTS STEERABLE
L – BAND	N.A.	N.A.	N.A.	N.A.	N.A.	EARTH COV.	N.A.	N.A.

Development of the INTELSAT Global system

Figure 30.4

The development of earth-stations has mirrored the development of the satellites and has necessitated large and expensive international/urban gateways connecting into a national telecommunications network. It is important to realise that all of this has resulted from particular service requirements within a particular network configuration.

New markets—the challenge

The global system is well developed and has made use of several new innovations in technology to cater for the increases in demand for capacity made upon it. Of particular current significance is the trend to digital communications in which the dominant speech traffic may significantly reduce its transmission requirements (bit rates) by the use of digital signal processing. The use of reduced-bit-rate speech for 64 Kb/s PCM to 32 Kb/s ADPCM and later to 16 Kb/s with appropriate coding, will allow a vast expansion in traffic with existing satellites. The introduction of a transatlantic optical fibre system (TAT-8) in 1988 and the resulting increase in capacity it will bring, together with the liberalisation on the transatlantic route, makes one really ponder 'wither satellite systems'.

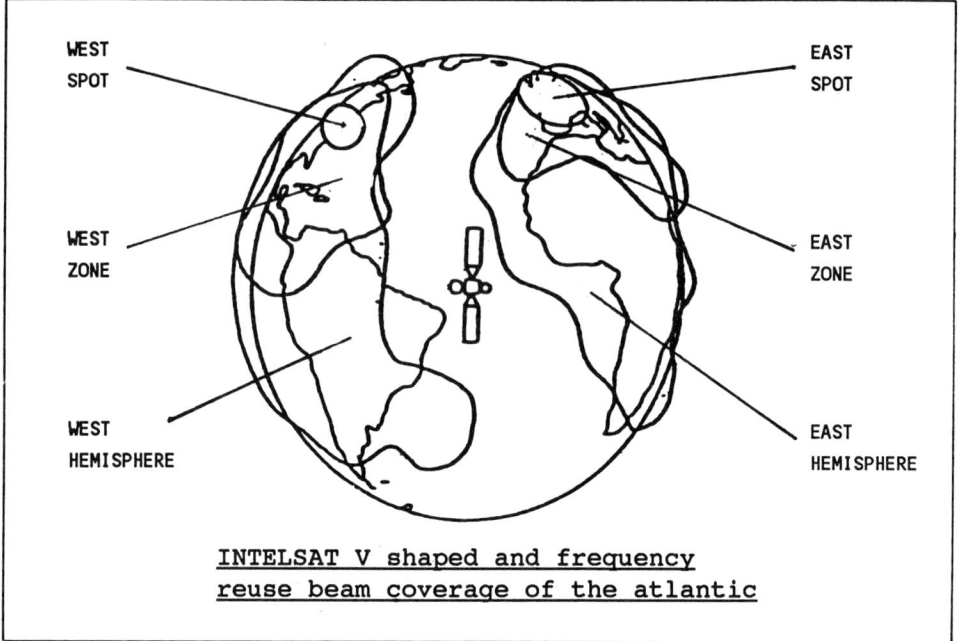

WEST SPOT   EAST SPOT

WEST ZONE   EAST ZONE

WEST HEMISPHERE   EAST HEMISPHERE

INTELSAT V shaped and frequency reuse beam coverage of the atlantic

Figure 30.5

I am convinced that we stand at a crossroads in the development of satellite communications, which is represented by the fourth era in Figure 30.2—the small dish/mobile era. These applications represent new markets to those traditional for our subject and require new solutions to new problems.

The nature of these new markets requires further examination. There are three new requirements which have already started to appear: direct broadcasting of television, small dish business systems, and mobile—maritime/aeronautical/land mobile.

The first of these, DBS, is already well advanced and higher power DBS satellites in the 1985 to 1990 timeframe become increasingly likely, broadcasting directly to the home. Although this is a potential money-spinner there are a few new problems for the satellite engineer to solve, although the challenge of producing a cheap and reliable domestic receiver for a mass market is already occupying the ingenuity of engineers. The success or failure of these systems is dependent upon the programme material and not primarily on the satellite technology. The average consumer does not care whether the programmes appear via satellite, cable or any other transmission medium. Thus no further mention of this application will be made in the discussion of future engineering challenges.

The real challenge to the satellite engineer lies in the other two markets, with either small fixed business earth-stations or various types of mobile system, for which the broadcast coverage of satellites provides a unique solution. Satellite systems operating to these new markets are already in existence today, for example via the global Inmarsat system to shipping, in the mobile area and the SBS system in the USA and ECS-SMS and Telecom I business services in Europe. The latter allows the use of three to five metre antennas which can be situated on the top of a building. However, the satellites currently in use are of the 'dumb' type illustrated in figure 30.6

These produce severe limitations for use in these new markets. In the small-station business systems they result in either excessively expensive earth-stations (£500,000 to £1 million) for TDMA

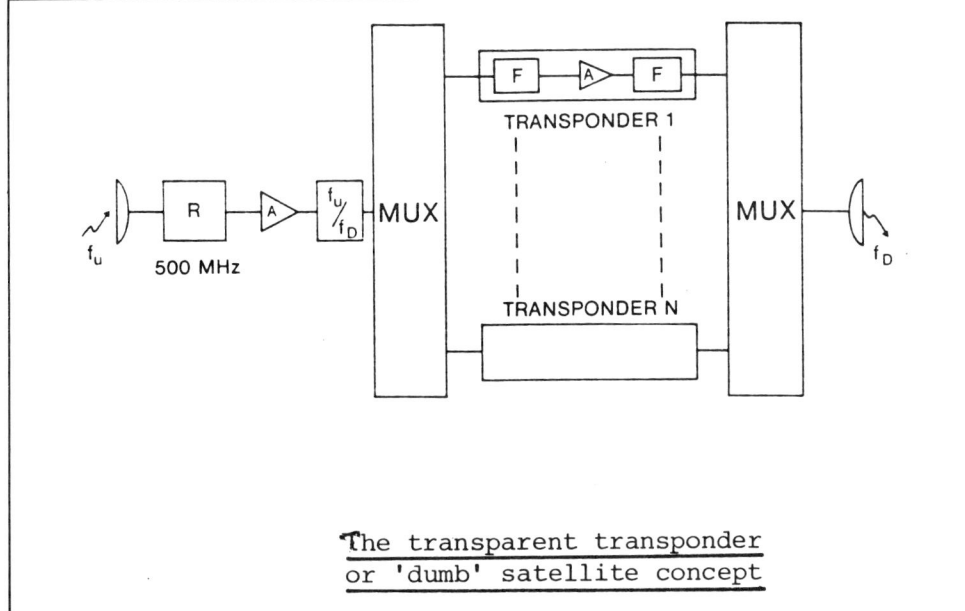

Figure 30.6

operation (Telecom I and SBS) or restricted capacity and hence crippling tariffs for FDMA (ECS-SMS). There is a proven demand for such systems, but current dumb satellites cannot satisfy it economically.

Figure 30.7 shows the demand for business services in Europe and their provision by means of various new satellite technologies: it is evident that new technologies are required just to meet the demands within the bandwidth available.

However, this is not the whole story. Although the technical development is a necessary precursor to the new applications, it is the economic and commercial environment which plays by far the largest part in determining what will actually happen. In this sphere the reduction of *both* the earth-segment *and* the space segment tariffs must be the driving forces. Whether the fixed service business or the mobile systems are considered, this means smaller and cheaper earth-stations together with higher utilisation of the space segment. These will only come about when 'the intelligent era', proposed in Figure 30.2, has been achieved.

New technologies

Our market-driven crossroads also leads to a technology crossroads, from 'dumb' to 'intelligent' satellites.

The following represent possible new technologies that could be used: improved modulation schemes, improved coding schemes, improved access techniques, new frequency bands—20/30 GHz, frequency re-use—spacial and polarisation, on-board processing.

Improved modulation and coding schemes are already in use and indeed the importance of speech coding in reducing the transmission requirements for this important service has been mentioned. As for modulation, simple BPSK or QPSK is currently in use in most of the digital systems. Although such systems are not spectrally efficient, there is currently enough bandwidth not to be concerned about it.

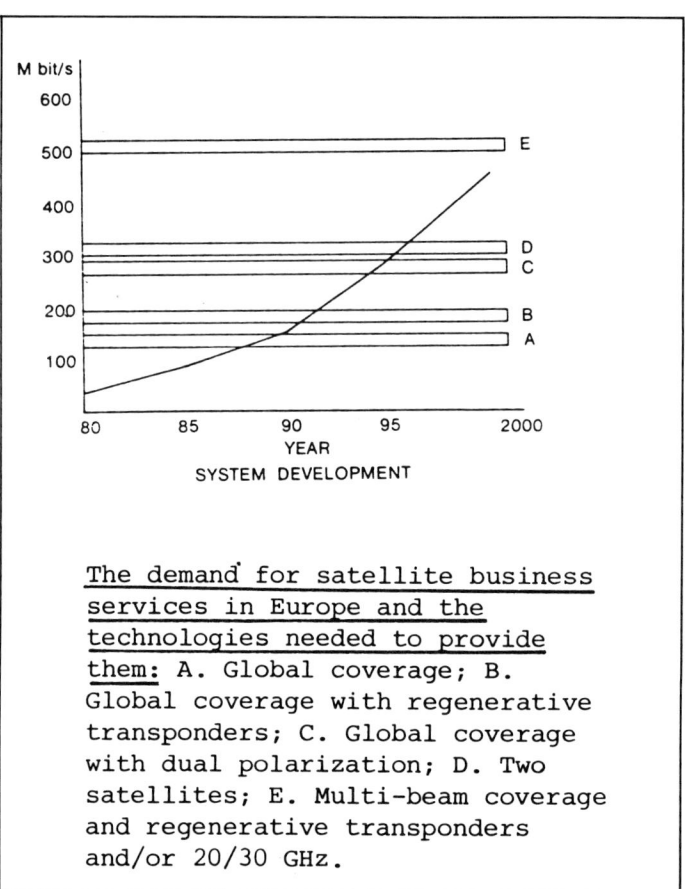

The demand for satellite business services in Europe and the technologies needed to provide them: A. Global coverage; B. Global coverage with regenerative transponders; C. Global coverage with dual polarization; D. Two satellites; E. Multi-beam coverage and regenerative transponders and/or 20/30 GHz.

Figure 30.7

However satellite systems are becoming more dominated by interference rather than the classical constraints of power or bandwidth. This may force future systems to both conserve bandwidth and provide resilient modulation schemes. However, neither of these techniques will completely solve the problem. Improved access techniques are coming closer to the heart of the problem. Currently used schemes have their drawbacks: FDMA suffers from poor satellite utilisation, but does enable cheap transmitting equipment; TDMA suffers from expensive earth-station equipment due to its burst-mode operation, but provides higher satellite utilisation. Various forms of spread spectrum (CDMA) can be shown to provide an economic solution for specialised systems, for example low-data rate transmission in a high interference environment (Intelsat's Intelnet service). However, satellite systems designers have not, to date, matched the access techniques to the services and have been constrained in doing so by the types of satellite available. This is clearly an area which could have potential advantages in the future.

The use of the 20/30 GHz frequency bands will open up much greater bandwidths but will produce two problems. Firstly, the rain fading at these frequencies is prohibitive and systems cannot be designed economically with a fixed fade-margin which may be 15 to 25 dB. Fade counter-measures are needed such as adaptive transmission and coding schemes, fade spreading or up-path power control. Secondly,

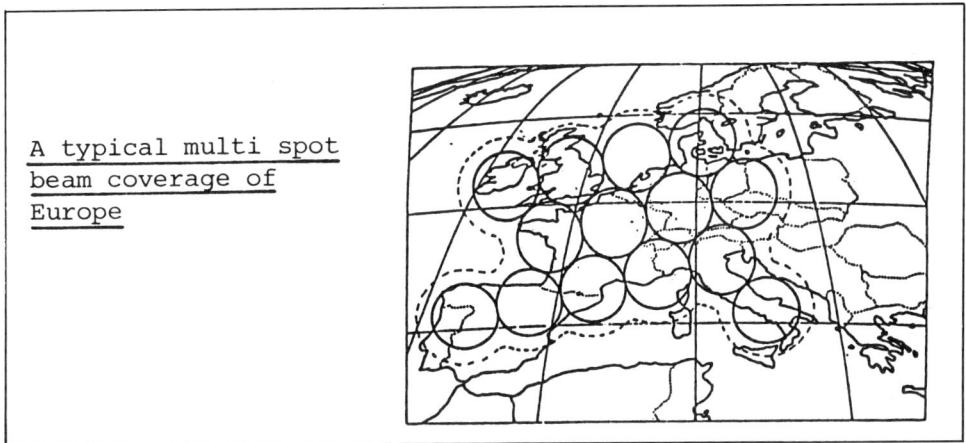

A typical multi spot beam coverage of Europe

Figure 30.8

earth-station equipment in the 20/30 GHz millimetre bands is, and will remain, expensive. Thus, unless interference in the other bands forces operators to move upwards, it is unlikely that these bands will be used for some time to come.

Frequency re-use has already been used to extend bandwidth at 4/6 GHz and 11/14 GHz using the two modes of orthogonal polarisations (this makes the earth-stations more expensive) and spacial separation. For some applications where the allocated bandwidth is plainly inadequate to meet traffic (maritime/aeronautical mobile) spacial frequency re-use will have to be used. It does offer another important benefit if it is extended to very small spot beams as shown in Figure 30.8 and that is the increase of effective power (eirp) which could reduce the size of the earth terminals. However multiple spot beams imply connectivity problems on board the satellite and this leads to the final technological innovation (and in the author's view the key one); on-board processing. This involves the inclusion of processors on board the satellite for the following functions: channel-to-beam routing (message routing), regenerative transponders (signal processing), and overall or partial network control (resource sharing).

The channel-to-beam routing can be performed at either r.f., i.f. or baseband. Baseband is preferred if regenerative transponders are included which enable the demodulation and decoding of the traffic on-board, thus breaking the binding of the up-link and down-link which is a feature of transparent transponders. This enables optimum modulation and coding to be chosen for each link and a saving in power made due to the on-board regeneration of the signals. On-board network control means that access schemes can be chosen to meet user needs and can be different on up- and down-links. Users with different transmission rates can be accommodated through the same system, enabling small and cheap earth-stations to be used by those with only small traffic requirements. On-board resource control means that each earth-station link can be allocated resources as and when it needs it and the system can be dynamically balanced as its use changes.

The intelligent birds

The satellites thus envisaged will include multi-processing and may take the form shown in Figure 30.9. Here we see the inclusion of multi-processors on-board the satellite. Processors will be the heart of the on-board demodulators and decoders as well as the new switches and on-board control. Individually they will need to be optimised to perform their major task and an architecture evolved for

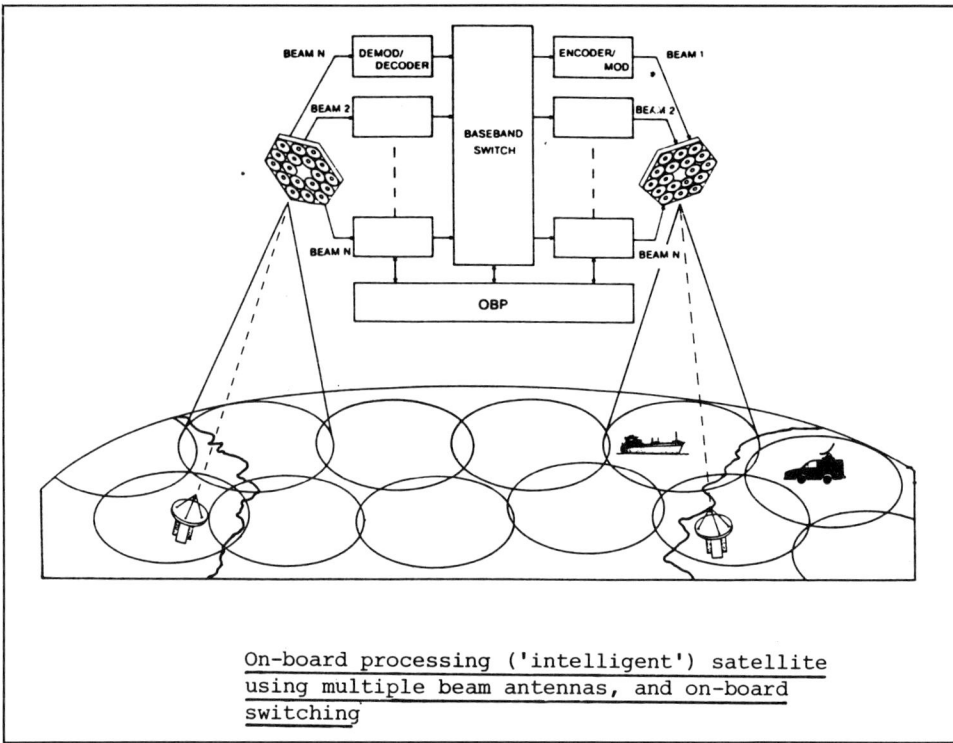

On-board processing ('intelligent') satellite using multiple beam antennas, and on-board switching

Figure 30.9

their interconnection and self checking. Security and fault-tolerance will be two major design considerations. A major part of the system will be software, and design to accommodate reconfiguration of the system, self testing and repair should all be possible.

The two markets which have been identified, small fixed business systems and mobiles, will both need to reduce the size, complexity and cost of the earth-terminals. The transmitter of a conventional earth-station is the largest single cost and this must be reduced by fitting it more exactly to the user needs. Most users require low-bit rate data communications and thus we must work with the cheapest access scheme commensurate with a low-power solid-state amplifier. SCPC/FDMA seems to fit the bill. In addition on the down link continuous wave TDM provides a receiver which is free of the complications of burst mode TDMA.

An on-board processing satellite with SCPC/FDMA up-link and TDM down, which can cope with variable rates from different users, message-switching and on-board control would appear to be the answer to the market demands. Such a design is shown in Figure 30.10 from which it will be seen that the two major new components needed are the T-S-T switch (the heart of a modern digital telephone exchange) and the time-to-frequency, frequency-to-time mapping device; a transmultiplexer which when accompained by demodulation becomes a multi-carrier demodulator. Both devices have been built terrestrially but in space new problems will need to be overcome.

These subsystems are comprised of processors, memory devices, data buffers and perhaps special processing devices. For any of the applications that have been mentioned, the demands are high in terms of both power dissipation, mass and volume of the on-board equipment. Obviously advances in the range of very large scale integration (VLSI) electronics and in new, high-speed, low-power-

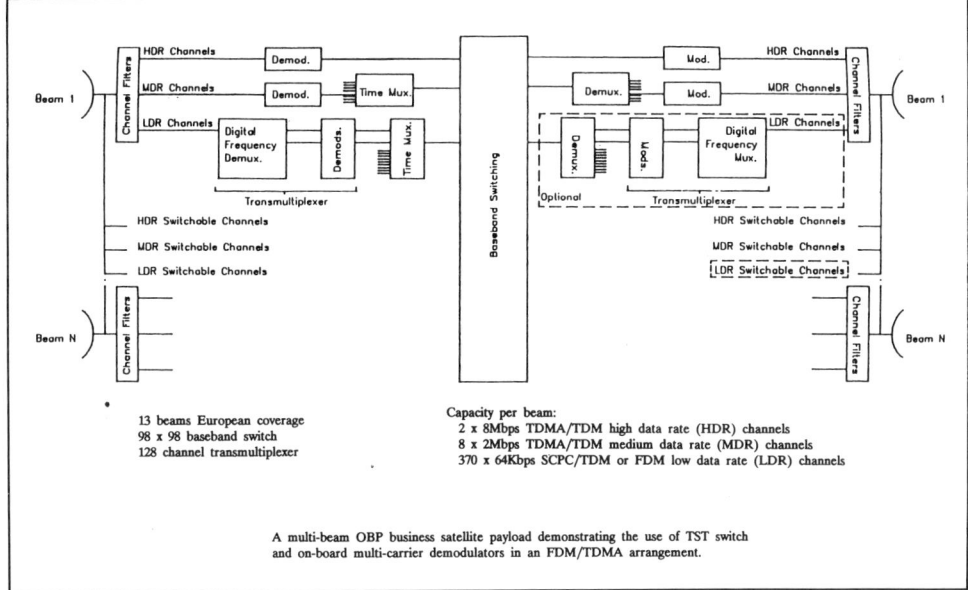

13 beams European coverage
98 x 98 baseband switch
128 channel transmultiplexer

Capacity per beam:
2 x 8Mbps TDMA/TDM high data rate (HDR) channels
8 x 2Mbps TDMA/TDM medium data rate (MDR) channels
370 x 64Kbps SCPC/TDM or FDM low data rate (LDR) channels

A multi-beam OBP business satellite payload demonstrating the use of TST switch and on-board multi-carrier demodulators in an FDM/TDMA arrangement.

Figure 30.10

dissipation technology, such as GaAs, will help. An additional problem is that of operation in a high radiation environment. This requires the technology to be radiation hard at the same time as having very low-power-dissipation; two characteristics which are usually in opposition, as evidenced in current CMOS technology.

Besides the device technology problems, considerable ingenuity in satellite design is required. Owing to increased satellite lifetime and potential changing roles during the lifetime, a reconfigurable and reprogrammable system is required. This has software implications, as does the overall reliability of a complex satellite. Fault-tolerant software and operating systems are required in order to improve the reliability of the satellites—there is still not much likelihood of a visit from the repair man! A careful balance between the implementation of the fault-tolerance and between hardware and software is in itself a whole new research area.

Will it happen? And when?

The scenario of the on-board processing 'intelligent satellite' presented in the last section is sometimes challenged. It has been explained that truly economic systems design, and therefore services at tariffs that the customer will be prepared to pay, is only possible with on-board processing. The protagonists usually advance the argument of lack of flexibility and reliability. The latter clearly has to be proven, and will be one of the major challenges to satellite engineers in the next decade. The flexibility argument is (in this author's opinion) not valid and borne out of too close an association with what has gone before. New markets will increasingly require dedicated satellites and this is the only way to make services cost-effective. Gone are the days of the all-purpose satellite in which the earth-segment was constrained to inefficiency—evidence business satellites of today! OBP satellites are being taken seriously. In the USA the NASA ACTS experiment has been given the go-ahead to demonstrate the interconnection of variable bit-rate users via an OBP satellite in the latter part of this decade. In Japan the ETS-V satellite will realise the SCPC/FDMA-TDM arrangement previously discussed in the same time frame. Research and development contraints are already let by both Intelsat (for the IBS service) and ESA (for the mobile services) for the development of the engineering models incorporating the new components within the next few years. Plans within ESA are to fly an experiment around 1995 of an OBP payload, possibly orientated towards the mobile services. It is generally accepted that a service via such a satellite will be in operation by the year 2000 for mobiles and that an Inmarsat third generation satellite system would incorporate such techniques. Within the UK the academic community has been leading the OBP research and has proposed a mobile systems demonstrator, originally CERS and now renamed T-STAT. This latter demonstrator will use the highly elliptic Molniya type orbit to provide overhead mobile services in northern and southern latitudes. Very recently (1987) ESA has recognised this work and has instigated its own highly elliptic orbit satellite, Archimedes. Although this does not contain OBP, the follow-on almost certainly will.

Hence despite the scepticism in some quarters, OBP will, in the author's opinion, happen. The technological problems will be solved by the end of the decade and operational satellites in existence by the mid-1990s. Now is perhaps the time to plan services via such satellites!

31

16 Channels and medium power: the logical way ahead

Marcus Bicknell

The limitations of satellite television today

The new media scene in Europe has in the last three years been dominated by the arrival of new satellite-distributed television channels crossing international boundaries. These new satellite channels are finding it difficult to attain profitability because of the small number of homes in which the current low-powered satellites can be received. Because of the high price of reception equipment and the large size of dish required to pick up the television channels from these low-powered satellites, the market so far has virtually been restricted to cable networks.

Eutelsat F-1

The satellite with the greatest European penetration for television homes is the low-powered Eutelsat Flight 1 which carries several important channels: Sky Channel, TV-5, RAI Uno, New World Channel, RTL Plus, World Net, Sat-1, 3-Sat, Film Net, Teleclub, Music Box/Superchannel, Europa TV (off the air since November 1986).

Despite the fact that Eutelsat F-1 provides an extremely European footprint, the low power of the satellite requires expensive reception equipment and dishes between 1.8 and 5 metres in diameter, which thus limits reception to cable networks.

Intelsat V

Intelsat V, at 27.5 degrees west, carries six very attractive channels which are all thematic. They respond well to the consumer's demand for all-day choice of programming and would be ideal on a satellite like Astra: Premiere (Pay-TV movie channel), Lifestyle, Arts Channel, Screen Sport, Children's Channel, Cable News Network (Ted Turner).

The potential market penetration for these channels is even less than for those on Eutelsat F-1 because the satellite's footprint is oriented towards the British Isles. The German Bundespost has had to take transponder capacity on another Intelsat V positioned at 60 degrees east, whose signals require very large dishes angled almost horizontally along the ground. The French PTT (Direction Generale des Telecommunications) has launched two low-powered satellites (Telecom 1A and 1B) which distribute national TV channel signals to re-broadcast transmitters round France.

All of these low-powered satellites require large reception dishes. Market growth has been limited and some of the other future satellite projects do not give much cause for hope of a breakthrough. The potential market for these new TV channels is therefore limited to 12.6 million homes connected to

AVAILABLE MARKETS (millions of households)

	Cable Homes (1)	MATV Homes (2)	Other Homes (3)	TOTAL TV Homes (4)
Austria	.3	.6	1.4	2.83
Belgium/Lux	3.1	.1	.3	3.62
Denmark	.2	.6	1.2	1.89
Finland	.2	.4	.9	1.62
France	.1	6.3	12.1	20.5
Ireland	.3	.1	.4	.8
Italy		5.8	11.6	18.4
Netherlands	3.6	.6	.1	4.6
Norway	.3	.3	.8	1.3
Portugal			2.9	2.9
Spain		2.5	7.4	12.3
Sweden	.3	1	1.9	3.6
Switzerland	1.3	.5	.1	2.3
U.K.	.2	2.5	17.8	20.8
W.Germany	2.7	7.0	11.5	24.5
TOTAL	12.6	28.3	70.4	119 million

(1) Home connected to wideband cable networks at the end of 1986. (Source: CIT Cable Report 1985 updated EPM 1987)

(2) The SMATV Market figures represent those homes in blocks of flats already receiving broadcast television via master antenna systems and therefore perfect for connection to new television channels. It does not include those blocks of apartments that MAY wish to subscribe to satellite delivered TV services when available. (Source J.Tydeman/E.J.Kelm "New Media in Europe" 1986)

(3) DTH (Direct To Home): other homes with TV sets except those passed by cable.

(4) All TV households in Western European countries.

Source J.Tydeman/E.J.Kelm "New Media in Europe", CIT and Mackintosh International, wideband cable updated by EPM January 1987.

Figure 31.1

cable in Europe, of which up to 8.5 million at the end of May 1987 were receiving one or more satellite channels. This is only 7.1 per cent of the 119 million homes in Western Europe with television sets. No wonder that advertising revenue is hard to come by.

The Astra service is a direct response to the clearly defined market demand for a satellite with more power. A satellite especially designed for television distribution and receivable on a small dish will give access to a much bigger market from Satellite Master Antenna Television (SMATV) and direct reception (Figure 31.1). There are over 28 million homes in blocks of flats receiving the broadcast channels through a community antenna. These households can be connected up to a community satellite antenna to convert the system into a SMATV outlet. Furthermore, there are another 70 million homes around Europe not passed by cable, of which 50 million have the consumer purchasing power to buy themselves individual reception equipment.

This provides a new market for satellite channels of 80 million homes, in addition to the 10 million homes already connected to cable. Even if only 10 per cent of this new market is achieved, then the new television channels will be able to double their existing subscriber base and, therefore, their advertising revenue. A partial success by Astra translates itself into profitability and survival.

SES considers cable the essential starting point for the programmers whilst SMATV and individual reception gives millions of other homes access to these new programmes. It is, in particular, the non-cabled homes and their rate of penetration by satellite television that are of interest to television channels already carried in cable networks via low powered satellites.

Forecasts of SMATV and direct reception growth

It has been estimated by independent research commissioned by SES (John Tydeman for EPM, January 1987) that a cautious assessment of market growth in cable, SMATV and direct reception,

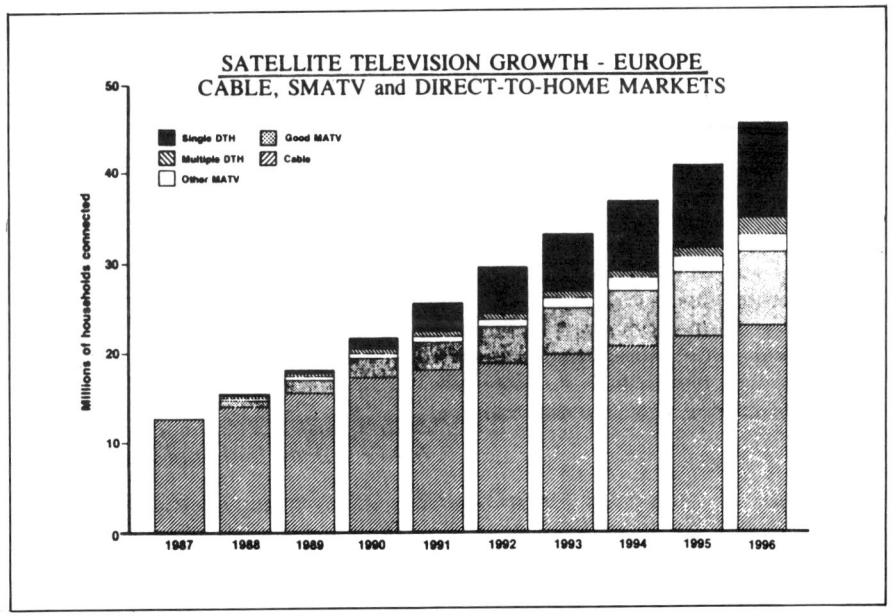

Figure 31.2

MARKET SUMMARY	ASTRA POTENTIAL HOUSEHOLDS:								
	1988	1989	1990	1991	1992	1993	1994	1995	1996
SMATV Homes									
Good MATV	0.6	1.2	1.8	2.7	3.6	4.5	5.4	6.3	7.2
Other MATV	0.1	0.2	0.3	0.6	0.8	1.0	1.2	1.5	1.7
Direct-to-Home Reception									
Single Dwellings	0.2	0.5	1.2	2.9	4.8	6.1	7.3	8.8	10.2
Multiple Dwellings	0.1	0.1	0.3	0.4	0.4	0.5	0.6	0.7	1.2
TOTAL SMATV & DIRECT	0.93	2.02	3.62	6.52	9.60	12.0	14.5	17.1	20.2
% of all TV Homes	1%	2%	3%	5%	8%	10%	12%	14%	17%
CABLE HOMES	14.0	15.4	17.3	18.1	18.9	19.8	20.7	21.8	23.0
% of all TV Homes	12%	13%	14%	15%	16%	16%	17%	18%	19%
TOTAL CABLE, SMATV, DIRECT	14.9	17.4	20.9	24.5	28.4	31.8	35.2	39.0	43.2
% of all TV Homes	12%	14%	17%	20%	24%	26%	29%	32%	36%

millions of connections

Figure 31.3

predicts 40 per cent of European television households (45.9 million) receiving television signals directly to home or indirectly (cable and SMATV) via satellites by 1996 (Figure 31.2). 43.2 million of these would be within the Astra footprint (Figure 31.3) of which 23 million will be receiving Astra via cable, and 20 million via SMATV or direct-to-home.

SES: the company

To trigger this mass market demand for satellite television, SES is now committed to launching Europe's first multi-channel, medium-power satellite. The company was created around a wide-based European shareholding structure. In addition to two State banks of the Grand Duchy of Luxembourg, there are two of the top five banks in West Germany, the Dresdner Bank and the Deutsche Bank. Other investors include two Scandinavian partners, Kinnevik from Sweden and Kirkbi, the Danish owners of Lego. There are four shareholders from Belgium and Luxembourg, including Electrafina who also hold shares in CLT/RTL. In late 1986 the Belgium State National Investment Corporation (SNI) invested in SES.

On Thursday, 8th January Thames Television, Britain's biggest commercial television station and programme producer, received the approval of the Independent Broadcasting Authority to take a five per cent stake in SES with options for up to 10 per cent. Thames's decision is a clear vote in favour of the medium-power 16-channel satellite option.

At the last capital increase in April, SES had $120 million available, over half of the eventual total of $180 million, half of which is equity capital and the rest loans. SES's funding has gone towards paying for the satellite itself, presently nearing completion in RCA's New Jersey plant. It is an RCA 4000 of a type in regular successful use round the world. Astra will be launched by Flight 27 of the European Ariane rocket, presently scheduled for June or September 1988.

SES's capital has also gone towards the building of Astra's ground station at Betzdorf near Luxembourg, completed in January 1987, where the Tracking, Telemetry and Control of the satellite will be handled. The idea becomes reality. Nothing will now stop SES putting Astra in orbit and providing the service which is the answer to all our commercial dreams.

Astra: Legal status

There were at the beginning some uncertainties about the legal framework in which the first private satellite in Europe would work. But much progress has been made. Although SES is a private company, Astra will operate under a franchise from the Luxembourg government for the use of frequencies and orbital positions. They have also been most supportive as 20 per cent shareholders, guarantors of bank loans and in their day-to-day work for the project.

There are also the recommendations of the Commission in Brussels and of the Council of Europe in Strasbourg, whose efforts at homogenising European media legislation will result in clearer guidelines for SES and the programmers. Conformity with applicable laws and collective agreements will be enforced on the programming and advertising aspects of the Astra channels under their contract with SES. The Fixed Service Satellite frequencies (11,200 to 11,450 MHz) and orbital position (19.2 degrees east) franchised to SES by the Luxembourg government completed in May 1987 the process of approval by the International Frequency Registration Board of the ITU in Geneva.

Although SES is installing the equipment necessary to uplink up to eight TV channels from Luxembourg, SES has the confirmation in Great Britain, and indications from several other European governments, that permission will be accorded for up-link of Astra's channels in their countries of origin. As for down-links, reception of Astra is already under the existing legislation of many European countries.

Although there has been some opposition to Astra from the existing European telecommunications satellite monopoly Eutelsat, SES and the Luxembourg government continue to try to find a way to cooperate with that organisation and now has the support of several of its PTT members and their governments. SES is keen to see a well-coordinated European space segment market because it is in the interests of the programme providers, the reception equipment industry and Europe's Television viewers.

Astra: The satellite

SES is the operating and holding company. Astra is the name by which the package of 16 television channels all receivable, within the 50 dBW area, on one small 85 cm dish, and its multiple sound channels for language versions, will be known to the public.

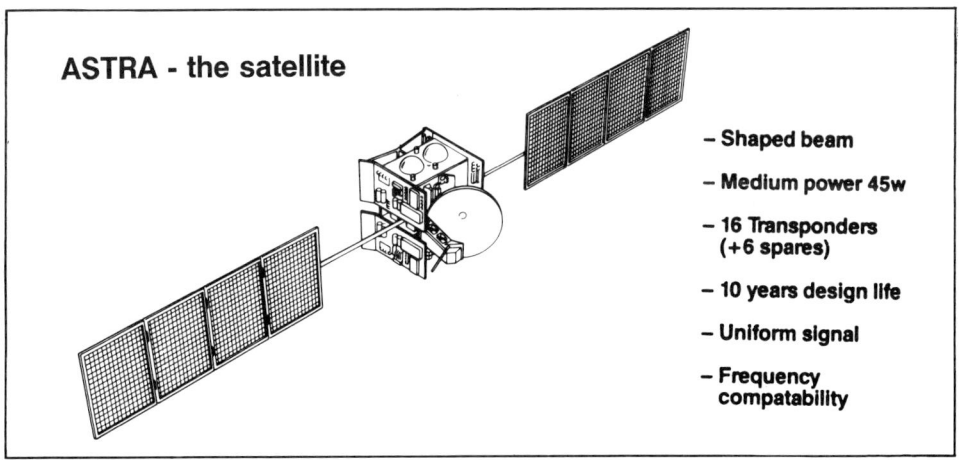

ASTRA - the satellite

- Shaped beam
- Medium power 45w
- 16 Transponders (+6 spares)
- 10 years design life
- Uniform signal
- Frequency compatability

Figure 31.4

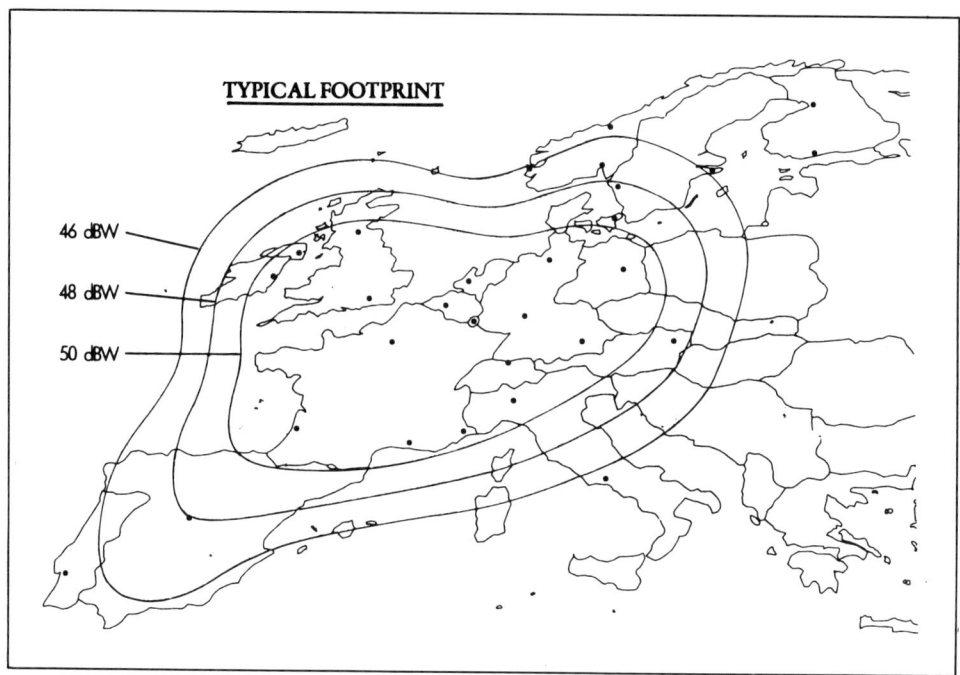

Figure 31.5

Figure 31.6

The essential commercial features of the satellite itself include a footprint specially designed for Europe, medium-power 45 watt transponders, 16 TV channels and six back-up transponders in case of breakdowns, a 10 year design life and a uniform down-link signal even in poor weather conditions. Unlike the proposed DBS satellites, Astra is equipped with batteries for eclipse protection; the transponders work 24 hours a day every day. Furthermore, Astra will use down-link frequencies in the middle of the Eutelsat Flight-1 and Intelsat V frequencies and uses the same linear polarisation parameters. This allows many existing items of reception equipment to be used for Astra.

One of the most advanced characteristics of the satellite is its specially-shaped footprint, as opposed to the oval footprints of Eutelsat F-1 and Intelsat V. Astra concentrates the television signals into those areas of Europe which are most highly populated. Therefore, 90 per cent of the whole consumer purchasing power of Western Europe is within the 50 dBW contour where the TV channels can be picked up on an 85 cm dish. Following the proposed changes to the Astra footprint, peaks of power will enable the quasi-DBS-sized 60 cm dish to be used in Great Britain and large parts of the Continent.

Although the 60 to 85 centimetre diameter dish for Astra is not as small as a Direct Broadcast Satellite dish (45 cms or less), it is aesthetic, small enough to hide in a corner of the garden or roof and small enough to bring home in the car.

The complete reception equipment for Astra, including dish, low noise amplifier and 16 channel set top tuner will in the future cost the same as a video cassette recorder. Indications from some individual aggressive manufacturers and from authorities like British Telecom International are that a complete set will cost as little as £350 pounds (DM 1,000 or FF 3,500) in PAL or SECAM or 15 to 20 per cent more if a MAC transmission standard microprocessor were to be built into the tuner. A dish size smaller than today and a vast increase in the size of the market for manufacturers should allow for further substantial equipment price reductions.

DBS satellites? Is there a market?

The low price of consumer reception equipment highlights the advantages held by the SES Astra system, a medium power satellite, not only over low power satellites limited to cable distribution, but also over the high power direct broadcast satellites to be launched by the French and German governments, and possibly by BSB, British franchise holder. These high power satellites will have three or four transponders at up to 230 watts each, giving high power but not much benefit to the consumer except for a slightly smaller dish. According to independent studies (BTI *et al*) DBS equipment will only be 8 to 12 per cent cheaper than Astra's if at all.

SES has chosen to ensure an attractive package of 16 channels on a dish which is only marginally bigger than DBS. The total package of equipment will be cheaper at the start because it exists on the market already, Astra's programme choice will be commercial not political: better choice for the TV viewer at lower cost. The viewing public will choose 16 channels on 60 or 85 cms rather than three or four channels on 45 or 60 centimetres.

Whereas DBS satellites have their footprint concentrated on one country, Astra's 50 dBW contour covers 77 per cent of European homes, 85 per cent of Europe's leisure spending power and close to 100 per cent of the main language markets including French, German and English. Programmers will be able to amortise their cost over a bigger market.

SES's trump card is the 16 channel package of programmes that Astra can offer the viewer. The 16 satellite transponders are for lease or sale to TV channel operators and SES will not directly participate in the programming of the channels. But the company realises the importance of assembling a package of programmes that viewers all over Europe will want to watch.

EUROPEAN MARKETS BY LANGUAGE AND BY LEISURE SPENDING

	Percent of Euro Leisure Spend	TV Homes within ASTRA 50 dBW	Leisure Percent Within 50 dBW
British Isles	16.4%	97%	15.91%
French	15.9%	99%	15.74%
German	33.4%	100%	33.40%
Scandinavia	10.2%	70%	7.14%
Italy	11.2%	49%	5.48%
Spain	4.7%	0%	0%
Benelux	7.1%	100%	7.10%
Other	1.1%	0%	0%
Europe	100%	77%	84.77%

Figure 31.7

Languages Spoken Across Europe
Percentage of Adults Speaking:

	English	French	German
1 Austria	25%	12%	100%
2 Belgium/Lux	26%	71%	22%
3 Denmark	51%	5%	48%
4 Finland	25%	5%	5%
5 France	26%	100%	11%
6 Ireland	99%	12%	2%
7 Italy	13%	27%	6%
8 Netherlands	50%	46%	41%
9 Norway	80%	10%	20%
10 Portugal	15%	10%	5%
11 Spain	20%	10%	5%
12 Sweden	80%	10%	20%
13 Switzerland	26%	55%	81%
14 UK	100%	16%	9%
15 W.Germany	30%	12%	100%

(Gallop Survey, 1985)

Figure 31.8

The new commercial satellite channels are simply chasing a commercial market and are most interested in the areas of highest leisure-spending, dominated by the 33 per cent German-speaking tranche of Europe's leisure spending.

The multi-language marketplace

Although present acceptance of foreign channels is limited there are significant numbers of Europeans who speak other languages, as the research results in Figure 31.8 indicate. But when confronted with a choice of domestic *and* foreign entertainment channels, most viewers will choose to be addressed in their own language. We can nonetheless expect that Astra's language-targetted channels, when accompanied by a big choice, will find success on an international level.

SES will ensure that the core services of SES Astra will be general entertainment television, advertising supported, each one targetted to a single language area. Based on client discussions currently under way, the British, French, and German-speaking markets will be the prime targets for Astra's programmers because of the concentration of consumer purchasing power there. Scandinavian services are also likely to be on the satellite.

In addition to general entertainment, SES Astra will carry several thematic channels, easier to dub or sub-title into several languages. The sports channel can carry major events for which the visual is more important than the commentary. Placido Domingo singing Verdi from La Scala on the Culture channel will be in Italian whatever the language of the viewer. These demands from SES's clients lead to a clear idea of the ideal line-up of channels on Astra.

The great attraction of this line-up to the consumer, and therefore to a programmer on Astra, is that at least eight channels will be of primary interest and a couple or more of secondary interest.

In the competitive environment for satellite operators in Europe for the next two years, the correct balance of language-targetted and cross-border channels will be critical to the success of the satellite and weighs heavily in favour of the multi-channel Astra option.

If a French-funded channel, for example, wants to exploit its programmes in dubbed form in Britain then TDF-1 will be useless. No English homes would equip for TDF-1's three French channels when the alternatives were BSB's three in English or Astra's eight or more in English. Therefore a DBS satellite with limited number of channels can only be considered a national or, at best, language area vehicle. These are also of course arguments in favour of a re-negotiation of WARC '77.

A 16-channel FSS satellite on the other hand has the capacity to offer three or four language-targetted channels in the markets comprising over 80 per cent of Europe's leisure-spending power. Astra will have German channels to ensure installation of reception equipment in German homes, enabling French and English-based channels to gain access to new markets. Even the frustrated French-based channel, suitably dubbed, could gain access to British and other European markets for the first time. We sincerely believe that this is the only formula that will enable new television channels, both pan-European and language-targetted, to get to profitability.

Consumer demand for choice

The consumer will surely be more motivated to buy or rent reception equipment if there is a choice of 16 channels from one satellite.

Whether Pan-European or regional, new television channels respond to the television viewer's proven demand for more choice of programmes. As in specialist magazines and many areas of our consumer society, markets become segmented by consumers who wish to buy information and goods in the format and packaging which addresses them personally. Whether thematic or general

AN IDEAL LINE-UP FOR A
16-CHANNEL EUROPEAN TELEVISION SATELLITE

A) EUROPEAN AND MULTI-LANGUAGE CHANNELS

1	24 hour News		Pan-European
2	24 hour Music	(1)	Pan-European
3	24 hour Sports		Pan-European
4	24 hour Culture	(2)	Pan-European
5	6h-15h Children's	(3)	Pan-European
6	Movies	($)	European

B) LANGUAGE TARGETTED CHANNELS

7	Entertainment & Info		U.K./European
8	Entertainment & Info		U.K./European
	15h-late Movies ($)	(3)	U.K.
9	Entertainment & Info		French
10	Entertainment & Info		French
11	Movies ($) or Thematic		French
	Thematic	(3)	French
12	Entertainment & Info		German
13	Entertainment & Info		German
14	Movies ($) or Thematic		German
	Thematic	(3)	German
15	Entertainment & Info		Scandinavian
16	Entertainment	French, German, UK or Scandinavian	

(1) "Music" refers to MTV/Music Box/KMP style

(2) "Culture" includes Lifestyle-type programming

(3) Children's or other specialist interest programming could share a transponder with movies or other evening programming as is already the case in Europe and USA.

($) "Pay-Television": encrypted programmes requiring the payment of a subscription fee.

Figure 31.9

entertainment in content, these new channels find favour with the public alongside existing national broadcast channels.

The ratings of the new channels prove that the consumer does want more choice of programming. In the United Kingdom, the research organisation JICCAR informs us that 24 per cent of programmes viewed in cabled homes are from satellite-distributed channels. This means that Sky Channel, in UK cabled homes, is watched by more viewers than BBC 2 and Channel 4.

The ideal line-up of programmes on Astra shows a number of thematic as well as general entertainment channels. The advent of a multi-channel satellite service enables us to give the TV viewer a wide choice of all-day one-subject thematic channels. Hit button 15 for sports, any time: 12 for music: 9 for a movie: 16 for the women's interest channel.

What are the advantages of thematic channel programming?

Multiple opportunities to view

If you miss a movie at 9.00 pm on a Saturday you will still be able to catch it at 7.00 pm next Monday. Choice of the most convenient time to view.

Ease of access

You are no longer tied to the programme schedule of your prime national channels. Switch on to the programme type of your choice at any time of day. Watch what you want to watch, when you want to watch it.

More choice for other family members

The father of a family has found peak evening viewing of the national channels pretty good for him. But what about the interests of women and of that important minority group, children? Now they can watch programming made for them at any other time of the day. Even with only one TV set in the home more individuals' TV demand is catered for.

Close association between viewer and channel

As with the reader of a specialist magazine, the viewer of a thematic channel is much more receptive to programmes packaged and presented exclusively for him or her, by programmers oriented towards the specialist subject, not to classic multi-subject broadcasters.

Advertising targetting

The receptivity of viewer to channel increases receptivity to advertising and sponsorship messages. By analysis of the channels reach and cost-per-thousand, the advertiser can be certain to get the 'rifle' message to more of his own potential buyers whilst paying less than the 'blunderbuss' general channel approach.

To help attract the existing and new channels which to go make up the ideal Astra line-up, SES has recently reached agreement with British Telecom International (BTI) to create a Joint Venture for the marketing of Astra channels in the UK.

The Joint Venture has the right to between seven and 11 channels depending on demand from clients, thereby ensuring that UK satellite TV channels, who access present satellites through BTI, can make arrangements to move to Astra at launch in legal and financial security. BTI is cooperating with SES in intensive marketing of Astra to programmers, reception equipment manufacturers and viewers.

Choice of transmission standards and encryption

Some of SES's potential clients can successfully run their channels only if a satellite-addressable encryption system is in place. The needs for encryption are at least 3-fold. Firstly, a typical movie service like Premiere needs to be able to encrypt its channel all day except for subscribers. Secondly, a mini-Pay channel like Screen Sport will want to be able to encrypt key programmes for a few hours a day and to offer them for a few pounds a month. Thirdly, a regional advertising-supported entertainment channel might acquire some programming on the basis that only viewers in the channel's region can see those programmes, Scan-Sat in Scandinavia for example.

SES policy on the issue of encryption is clear.

> The number of encrypted channels on Astra, Pay-TV movie services or other, will be limited in number (say three to five) in order to give a maximum choice of free-of-charge channels to a TV viewer who buys the equipment.

> As Astra is transparent to most encryption systems, the programming clients of Astra will decide which system to adopt but all will be influenced to decide on the same system so that one viewer can access all 16 channels.

> The likelihood of multi-lingual sound carriers, possibly in stereo, for Astra's cross-border services in the long term argues in favour of a system with multiple stereo sound carriers.

> Several clients have expressed the intention to choose a state-of-the-art satellite-addressable conditional access system which controls the subscribers channels individually by a chip set in the receiver and with central computer control for central billing. SES is particpating at all possible levels in the discussions on these topics and will be supportive of any cost-effective systems needed by the programmers.

SES hopes in any case that a decision will be taken as quickly as possible, based primarily on the commercial feasibility of the system and the needs of the programmers, influenced by political factors only if essential to the success of the programme channels on Astra.

Marketing and services

Astra is not just a satellite. It is a service to television viewers, television programmers, and reception equipment manufacturers. For the first time outside of cable television marketing, we are jointly going out to sell a package of television programmes to the general public. They will not just automatically buy the package. We must sell it.

The success of ASTRA as the Hot Bird for European television distribution depends on the degree to which the 16 programme providers can succeed in penetrating the markets described above. To trigger the installation of receiving equipment in millions of homes, the marketing of the channels must be well-planned and intensive. SES will work in parallel with the 16 programme providers and equipment manufacturers to help them maximise their penetration.

As a background to this marketing strategy, it is assumed that Astra channels will be taken on cable networks whose operators decide in favour and for whom the cable operators undertake the marketing. We also assume that flat-dwellers will either have access to simple methods of SMATV networking or will be able to install personal equipment. But these flat dwellers (and their decision-makers) need awakening to the attraction of receiving Astra channels almost as much as the individual direct-to-home reception public. To penetrate this most elusive and most rewarding of the market type, the non-cabled homes, we must develop, jointly between the reception equipment industry, satellite operator and TV channel, the marketing strategy necessary.

SES's support can be divided into three essential categories, client support, reception equipment promotion and consumer marketing.

Client support

This entails the advance cooperation which SES can provide in helping the programme provider to prepare for launch of his TV service and to analyse results. SES will make available certain optional services aimed at helping the programme provider to prepare for launch of his TV channel and to analyse results. SES's extensive work on its own account puts the company in a position to offer advice on regulatory issues. SES technical staff are on hand to give technical advice and recommendations on transmission parameters. SES has started work with collective antenna installers to develop simpler and more cost-effective ways of connecting up blocks of flats.

SES will make available market analysis and research information, such as John Tydeman market predictions, to clients. SES is also closely involved in industry organisations and in several initiatives in market research in individual countries and on a European level. Amongst the tangible services already announced are the facility to up-link a channel from Luxembourg, and other technical services such as portable up-link, equipment testing and TV service monitoring.

Reception equipment promotion

SES has introduced, with the industry's support, a campaign running from January 1987 until launch and afterwards, on the reception equipment side. The right equipment must be available in retail at the right time and at the right price. SES will not manufacture or distribute equipment, nor propose any exclusive arrangements with one manufacturer or another. It is in the interests of the programmers, the TV viewers and the whole reception equipment industry that the help SES gives is generally available. All the effort will go into increasing overall market penetration.

To complement the consumer marketing effort of the channels and of Astra, SES will work very closely, especially prior to launch, with the hardware manufacturing industry to ensure that attractively-priced reception equipment is easily available to a potential buyer asking how to get Astra.

Good information would seem to be one of the key needs. Many dish buyers would normally, even in a year's time, be bewildered by the technology, by the choice of satellites and future transmission parameters. Therefore the industry should work along the same lines and unite to communicate accurately its commercial proposition to the consumer: help the consumer understand that the dish installation will be no more complicated than having a broadcast antenna installed.

One key element of this promotion, therefore, is the help given to retail outlets in identifying to a potential buyer that a particular piece of equipment is suitable for the reception of Astra. The operation will trigger a response from the retailer towards the manufacturers and distributors and will encourage the retailer both to stock the equipment and to carry displays and promotions in the store. Several key manufacturers have already lent their support and advice to the operation. A first phase entails the simple sticker concept. This will ensure that distributors of elements of satellite reception equipment which are compatible with Astra will be authorised to use the triangular sticker or label indicating 'compatible with Astra'. SES and the industry will actively promote the fact that many dishes, low-noise converters, connections and indoor tuners are already available on the European market in the frequency range and polarisation characteristics of Astra. This is because Astra uses the Fixed Service Frequency band like Eutelsat and Intelsat. The advantage is clear for a purchaser of an

LNC or a receiver for an existing satellite . . . the equipment will also serve him for ASTRA when it goes up.

Although the idea is simple in the case of dishes and LNC's, we can anticipate problems of transmission parameters with the indoor units or tuner/receivers. A tuner set up for PAL is not going to be able to handle D2-MAC without upgrading. But SES's proposal is that a tuner for which upgrading to MAC is provided (for example, by the slot necessary for the later insertion of the chip card) qualifies for compatibility.

As the ultimate service to the consumer, key distributors plan to market complete sets of Direct-To-Home reception equipment, including antenna with fixed mount, low noise converter, cables and connectors, 16-channel tuner/demodulator with infra-red remote selection. Such a package will be ideal for maximising penetration in high street retail chains like Darty, FNAC, Neckermann, W.H. Smith, Granada Rentals, Dixon's and Lasky's for whom stocking multiple elements would be less desirable.

This visual aspects of Astra's promotion will be useful to distributors and retailers. SES is willing to discuss making posters, displays, promotion video-cassette masters and other information material available to those manufacturers who choose to work together with SES and the programmers.

In the same way, SES will feature equipment manufacturers in their consumer advertising, and manufacturers will mention Astra on their publicity. Joint industry marketing plans country by country, perhaps with the coordination of industry groups like the Bundesverband Kabel und Satellit (West Germany), Simavelec (France), and the European Satellite Television Association, are envisaged.

SES is also cooperating with key retail outlets on the training necessary, probably by regional half-day seminars, for their store salesmen so that they can inform potential dish buyers of the choices open to them.

Consumer marketing

Astra recognises the need to work with television programmers to trigger consumer demand in the new Master Antenna and Individual Reception markets. Each channel will be promoting, independently from SES, its own service in order to attract viewers and to maximise advertising or subscription revenue. But the uniqueness of Astra's 16-channel choice calls for coordination between those campaigns and for a supplementary Astra campaign. SES has declared its intention to commit funds in those territories where the key Astra channel operators are in favour of the Astra campaign.

The Astra campaign commences two months before the launch of Astra, whilst the most significant purchases of advertising space will be confirmed within hours of the successful launch of the satellite. As Operational Service Date (start of television transmissions) will be 30 to 45 days after launch, press space and poster advertising will coincide with the channel launches.

The campaign is designed to trigger the consumer's recognition of the Astra satellite system, in particular that 16 new television channels are available on small dishes from a point in space, Astra. This was one of the key reasons why Astra was given a consumer-oriented name rather than the initials of the company name, SES, which are difficult to retain and to associate with. SES intends to target the advertising at non-cable areas in order to maximise its effect and to avoid disturbing the cable operator's own marketing.

Futhermore television and radio advertising is unlikely to be cost-effective for Astra. So the media retained by SES's study group, which has included key clients and TV programmers, will be concentrated on press, TV guides and display advertising.

The potential consumer will be led down the following path. The typical bored family, searching in vain for something good to watch will be hit first by the advertisement in the regular TV guide. Display advertising in newspapers, and on outdoor posters leads the potential buyer to an equipment outlet, whether it be a department store, electronics shop, TV antenna installer, specialist satellite TV company or TV rental branch.

The potential buyer is struck by the window display for Astra in the local shop (in some cases with video cassette recorder promoting Astra or a live feed off the satellite) and by the *Astra Compatible* stickers and tags on compatible equipment. Hopefully he is struck by the good advice given by the sales staff. We can all help him by information videos and training aids.

As soon as an individual enquires about Astra he or she could be given free of charge, or at subsidised cost, a *Guide to Satellite TV*, which will give simple information on choice of all TV satellites and channels, not just Astra, the legal and technical requirements and background information.

The enquirer is given a small informative colour pamphlet devoted to Astra and the channels it offers, so that in the family circle, in which the equipment purchase decision is often made, individual members of the family can see what channels will appeal most. Music and Children's for the kids? Lifestyle and Culture for mum? Sports and News for dad for the rare moments he's home?

Assuming the purchase is made and the equipment installed, the marketing of this happy family need not stop there. Each purchases of Astra equipment could be given one year's subscription to the Satellite Television Times, or an equivalent programme guide of satellite delivered television channels available on Astra and other key satellites. The subscription records could be used as a base for computerised billing information later. But, more importantly, if the client discovers other interesting channels on Astra through the guide then his satisfaction with the product will spread by word of mouth to other potential buyers. Fashion is the cheapest and most effective marketing device of any sort.

Many satellite TV programmers and hardware manufacturers believe Astra is the satellite most likely to develop the market. The 16-channel package and well-coordinated marketing will be vital in triggering a positive consumer response and the installation of reception equipment in tens of millions of homes across Europe.

32
DBS: understanding the risks and opportunities

Jon Chaplin

Introduction

The current problems with satellite launch systems must be seen in perspective and not distract attention from a number of areas relevant to this paper. The importance of space and satellite communications in particular means that Europe and the USA will overcome the recent setbacks with Ariane and the other expendable launchers. The future of the Space Shuttle for commercial satellite launching is less certain and in any case is unlikely to make much of a contribution before the end of this decade.

The first four high-power European direct broadcast satellite (DBS) systems will be launched before the end of this decade. This new means of carrying TV programmes, information and services to homes and small businesses should be seen as offering opportunities to many. At the same time the risks associated with such 'green field' ventures must also be appreciated by service providers, receiver manufacturers and the satellite industry itself.

Who will these service providers be? Established broadcasters, telecommunications administrations and print publishers are obvious answers but will there be new companies emerging that do not fit properly into any of these categories? They have been called 'New Media' companies but what does this mean?

The opportunities

Opportunities come from the possibility of reaching all in Europe, with a deeper penetration and faster coverage than other means including the enigmatic Integrated Service Digital Networks (ISDN). Wide international coverage enables fragmented markets to be reached by a new telecommunications means which only the telephone could do before. There are innumerable opportunities for European collaboration to exploit these new markets. Such ventures could be simply between two countries or whatever number it takes to reach the necessary level of resources. It seems unlikely that DBS overspill can be used to conduct business in another country without the support of a resident organisation.

All sorts of new services can be foreseen. Entertainment, particularly TV, is what many think of as DBS. Undoubtedly in the long-term DBS will carry many national and international TV services, including many that exist now and are carried by terrestrial networks. But the growth of such services is limited ultimately by disposable income for leisure and, more importantly, viewing time available. Many other types of service, classified broadly as information, will emerge and grow steadily without these constraints.

If an increasing amount of leisure time is going to be available to most of us, the demand for education will increase and many of us will be able to achieve fulfilment as well as some improvements in our job prospects. Distance-learning actually at home is seen by many in the field of education to be an important growth area. But education is just an example. Any information that we get from the words printed on paper could be distributed electronically to our own advantage. The main difficulties come in the storage, display and reproduction of this information.

These problems will probably first be solved for the small-business market. DBS is an ideal way to distribute information to the vast number of such enterprises that provide most of the new jobs in our society. The economics of lower power telecommunication satellites makes them unsuitable and the terrestrial networks will either be limited in their geographical coverage and/or bandwidth. Thus DBS could make a direct impact on European problems of economic development and job creation. Only some imagination from potential service providers is required and this includes government itself.

The risks

Dropping the satellite into the water after launch is but one risk, but is must be emphasised that the long term record for launchings is over 90 per cent success based on about 400 launches made over the last 20 years with six different launcher designs. Within the next three years Ariane should show that she can improve from the present 78 per cent success rate to match the 90 per cent of the most established American launchers. Experience has also shown that the chance of satellite failure after reaching geostationary orbit is even less, at a few per cent. So the temporary problems here are really those of an adult industry which has regressed into teenage adolescence and all that that implies. The result has been that satellite insurance is very difficult, like that of teenage motor cyclists. This has a knock-on effect on the commercial financing of some satellite systems.

It should also be mentioned that satellite and launch failure has disastrous consequences only for those systems relying on the successful deployment of a new type of satellite. This, of course, is exactly the situation with DBS at present, although a number of new services have been started with telecommunication satellites. Here the problem is less crucial being limited to the date when the more performant DBS satellite becomes available. This strategy obviously has advantages in security of service.

There is risk in DBS at this moment since it is essentially a 'green field' operation. The problem relates mainly to the provision of receivers for the public and the rate at which these will be installed in homes. DBS receiver manufacturers tend to play down their preparations for a new domestic market due to the competitive situation. There is still some uncertainty regarding the precise compatibility of the various members of the MAC-packet family and which standard which manufacturer will offer. However I do not believe that receiver availability will be the pacing item.

The big problem for all potential DBS service operators is the rate of growth of revenue. The costs of providing the service are known in advance and obviously all steps to minimise these costs are necessary to obtain up-front financing. But the crucial thing must be to maximise the probability of realising the predicted growth of revenue. This will be achieved if a sufficient variety of services is available in the home via DBS and that as many homes as possible can obtain a service if they wish.

Thus I have difficulty for example with a British three-channel service which is not complemented by channels on an Irish satellite and/or not receivable for some reason by a number of important niche markets in continental Europe. There are also a number of information services that could be provided by a DBS system and generate significant revenue in the early years (when programme services may in any case be filling the channel less than later on). A good example is closed-user-group video services, perhaps for a business. Regulation prevents such a use being made of a DBS channel but this will rob a

DBS operation of one of its chances to survive. There is no suggestion that business services should be forever carried by DBS nor that they will impede the development of publicly available services. But in the end an unknown rate of take-up of these services, and the resulting revenue growth, can make it so difficult to judge their viability that the project is abandoned as was the case with British DBS in 1984.

Other risk factors are regulation and state monopoly. This is seen as a major area for change but how quickly will it come? This has been the subject of a European Commission report designed to promote debate but the British have a tendency to think that Europe is a place to go for holidays not to do business. At least with DBS reception there are not the uncertainties associated with regulation of reception and monopoly restrictions implicit in the use of telecommunications satellites for distribution.

The 'bottom line'

In summary, the opportunities offered by DBS are unusual as well as considerable. Naturally there are risks but these can be measured in some cases and minimised in all. In any new enterprise there is an element of risk and it is the successful entrepreneur who decides to accept that risk. Ten years from now, I anticipate that DBS will be moving into a key and unique position in our society.

Appendix A

About the authors

BENSCHECK

Wolfgang Benscheck holds a university degree in business administration. Since 1979 he has been engaged in electronic publishing as managing director of Hoppenstedt Wirtschaftsdatenbank, a subsidiary of Verlag Hoppenstedt & Co. Prior to this appointment he was working in financial and credit analysis.

BICKNELL

Marcus Bicknell has, since April 1986, been Commercial Director of Société Européenne des Satellites, the company launching Europe's first private satellite, Astra, in 1988. After 15 years in the record industry with CBS and A&M Records, he joined Thorn EMI in 1983 as Head of Marketing for Premiere, the Children's Channel and Music Box. After a year he concentrated entirely on network and marketing for Music Box, Europe's all-day music television channel, achieving distribution via satellite into 4.5 million homes in 13 countries.

BORNSTAEDT

Falk von Bornstaedt studied physics at Old Dominion University (Norfolk, Virginia, USA) and economics at the universities of Bonn and Paris. He received his Diplome de licence d'economie appliquée in Paris in 1977, and Diplom-Volkswirt in Bonn in 1980. He has been chairman of Studiengruppe Bildschirmtext eV since 1981. In 1983 he was appointed a member of the editorial board of *Net* (a monthly journal), and is editor of a book series (Btx-Reihe) with 14 titles on Bildschirmtext. He is a member of the Institute of Applied Information Technology of GMD in Bonn.

BRUCE

Dr Margaret Bruce is a lecturer in management sciences at the University of Manchester Institute of Science & Technology. Her research and teaching interests focus on the management of design and innovation, strategic marketing and competitiveness, particularly with regard to the information technology industries.

BURNS

Christopher Burns is president of Christopher Burns Inc, in Salem, Massachusetts, a research and development firm that concentrates on the information industry. Prior to starting that firm in 1983, he was Senior Vice President of the Minneapolis Star and Tribune, Vice President for the Washington Post Company and a senior consultant at Arthur D Little Inc, where he directed much of that firm's work in future information systems.

CHANDOR

Anthony Chandor is Chairman of Mandarin Communications Limited, a consultancy specialising in information systems and electronic publishing. He recently lead a study tour of the US optical storage industry on behalf of the UK Department of Trade & Industry.

CHAPLIN

Jon Chaplin has been in satellite communications for 20 years and has spent most of the last ten years with the European Space Agency. In this period he has been concerned with all the satellite broadcasting activities of the Agency. He sees satellite communications as an ideal way of distributing information of all kinds.

COATES

Employed by DEC in many capacities since 1970, Eric Coates has specialised in providing market support for the CD-ROM over the past year. An Oxford arts graduate with 15 years' computer experience, he finds this variegated background useful for exploring the byways of the CD-ROM business.

EVANS

Barry G Evans is currently Alec Harley Reeves Professor of Information Systems at the University of Surrey, where he heads the satellite and telecommunications research activities. He has consulted worldwide on satellite communications systems and has been responsible for pioneering several new techniques in satellite systems. He is editor of the *International Journal on Satellite Communications*.

FERBER

Leon A Ferber is Executive Vice President of Perception Technology and co-founder of the company. He has nearly 20 years' experience in speech processing technologies and holds three patents for inventions. Raised and educated in Israel, he holds a BS in electrical engineering from Northeastern University, (Boston, Massachusetts, USA). Prior to establishing Perception, Mr Ferber was with Digital Equipment Corporation as a senior development engineer.

FINNIGAN

Paul F Finnigan has 30 years' experience in the design of computers and telecommunications hardware and software. Chairman and founder of Voicemail International of Santa Clara, California, Finnigan conceived and designed Dow Phone, one of the first commercial products utilising the audiotex capabilities and developed the crew scheduling application for TWA and automated reservation service for PSA.

FINZI

Claude Finzi is project manager at Télésystèms Questel for large international projects. He is President of the GIE ScanEurope.

FLEURY

Lionel Fleury is the general manager of Polycom, a subsidiary of France Cables et Radio and Agence France-Presse (AFT). Polycom has operated a satellite data broadcast system over Europe and America since 1987. A graduate of Ecole Nationale d'Administration (ENA) with a PhD in geophysics, he also heads a seminar on telecommunications economy to PhD students of Paris-Dauphine University. He was marketing director of the Telecom I satellite programme in the French PTT (1979–1983) and Chairman of the European Videoconferencing Experiment (EVE) in the CEPT from 1980 to 1983.

GAFFNER

Haines B Gaffner is president and founder of LINK Resources, the leading market research and management consulting organisation specialising entirely in the new electronic media. Mr Gaffner has 23 year's experience in the information industry, serving as Vice President of Business International and Quantum Science, and co-founder of FIND/SVP. He specialises in assisting executives to clarify their strategic aims in the electronic services field.

HEIDRICH

Wolfgang Heidrich, Dipl Ing, has been Director of the Bildshirmtext Service since 1984. Previously he was Head of the Bildschirmtext System Technology Section and was, in that capacity, responsible for the development of the new Btx-System. At the international level, Mr Heidrich is Chairman of the Videotex Rapporteurs Groups of CCITT COM I and CEPT.

HOWARD

John R Howard is manager of the central research team of Plessey Major Systems Ltd. He is responsible for a number of telecommunications research activities including RACE broadband switching and customer access connection. He is also project manager of the recently announced Eureka project to develop a broadband system.

KERR

Gordon W Kerr, MA CEng MIEE FSCTE, joined BT Research in 1975. In December 1981 he was given the responsibility to define, develop and initiate the video library service for the BT switched-star network now being installed in the Westminster franchise. He has also worked on some committees of the National Interactive Video Centre. Gordon Kerr was elected a Fellow of the Society of Cable Television Engineers in April 1986.

KLEYN

Howard Kleyn's professional career in Cable & Wireless plc focused on network design and computer-controlled message and data switching. His overseas experience includes the presidency of the Hong Kong Computer Society and the chairmanship of a colonial committee on data privacy. At one time he headed the Group's advanced development facility at Essex University and he is the author of a book on the world's data services. Having participated in the inception of the new UK PTO, Mercury Communications Ltd, Howard was later appointed Director, Corporate Strategy. At the beginning of 1987 he took up the post of Deputy Chairman of Oyston Cable Ltd, a cable television multiple systems operator.

KUSHNICK

Bruce Kushnick founded National TeleVoice, a consulting firm specialising in interactive voice services.

MAURER

Professor Hermann Maurer is Head of IIG with some 45 staff members, over 30 of them working in videotex-related research and development. He is also Adjunct Professor at the University of Denver USA, and advisor to the Austrian PTT in the area of telematics. Maurer's current interests are videotex and applications thereof, particularly in the area of computer assisted instruction. Professor Maurer is author of over 200 papers and four books.

RAUH

Thomas R Rauh is a Management Consulting Partner and National Services Director for Retail Consulting, based in the San Francisco office of Touche Ross. He has consulted to retailers both in the United States and abroad in the areas of strategic planning, financial planning and control, systems planning and project management, operations improvement and organisation development. Mr Rauh directs the firm's consulting practice in electronic shopping including projects for system operators, hardware and software vendors, retailers and consumer product manufacturers.

REED

Dr C William Reed has analysed new communications services for LINK Resources Corporation for the past three years.

ROBINSON

Alan S M Robinson is Executive Chairman of The Broadband Commications Company and Chief Executive of West Midlands Cable Communications Ltd, a company established in 1987 to obtain the cable telecommunications franchise for Birmingham and the West Midlands (468,000 homes). Mr Robinson was founder and Managing Director of Croydon Cable Television Ltd and ran that system for four years, having first become involved in the cable television industry in 1973. Mr Robinson trained in physics and aerospace engineering and has a BA degree from the University of Texas, having lived in the United States for some 14 years.

RUMBLE

David Rumble joined PA Computers & Telecommunications from British Telecom, where he was Head of Market Planning and Publicity for what is now the Business Services Division. His work at PActel includes a leading role in the development of scenario analysis methods for the CEC's RACE programme.

SEDDON

Graham Seddon is a co-founder of BRS Europe, a UK company which markets and supports software products developed by BRS Information Technologies, a leading vendor of on-line information. BRS Europe has been involved in CD-ROM development since 1985; its software is used in Whitaker's BOOKBANK product, launched in 1987.

SHORROCK

David Shorrock is a Senior Consultant, specialising in communications, with Logica, a leading independent computer software, systems and consultancy company with activities worldwide. David has over 12 years' experience in the communications and IT industries, working for various clients in government, public and private industry and telecommunication authorities. He regularly speaks at international seminars and conferences and has published a number of papers in international journals.

SUGIMOTO

Michio Sugimoto is senior manager of CAPTAIN systems in the Engineering Department of NTT. He is engaged in the planning and development of the CAPTAIN system. He is a graduate of Waseda University and has worked in the data communications services since 1972.

TERASHI

Toshio Terashi is associate manager of CAPTAIN systems in the Engineering Department of NTT. His is engaged in the planning and development of the videotex communications network. His is a graduate of Osaka University, and has worked in the data communications services since 1977.

TURNER

Byron M Turner is president of European Interactive Media. Prior to joining the company he served for over two years as Director of Creative Development in Europe for Activision Inc, responsible for establishing Activision's home computer software operations in Europe. Turner also spent 11 years with Thorn EMI in a variety of positions, most recently as director of Thorn EMI Video.

Appendix B

References

1 Introduction

1 R E Rice, *New Media Technology*. The New Media, Communication, Research & Technology, Ronald E Rice & Associates, page 35. Sage Publications, 1984.

2 *New Media and the advanced information society*, Report No 58, pages 1–35. JIPDEC, Japan, 1984.

3 R E Rice, *Development of New Media Research*. Page 16, *Ibid*.

4 T Namekawa, *Technologies on New Media in Japan*. Telecommunication Technologies, pages 94–96. North Holland, 1984.

2 Global markets for videotex services

1 M Tyler, *Videotext, Prestel and Teletext: The Economics and Politics of Some Electronic Publishing Media*. Telecommunications Policy, Volume 3 Number 1, pages 37–51. 1979.

2 *Vive le Teletel*. Systems International, Volume 15 Number 11, pages 97–100. November 1987.

6 Implementation of interactive videotex

1 *Bildschirmtext*. Institut fur Bildschirmtext, January 1988.

2 *Videotex International*. Proceedings of the conference, December 1986. Online, 1987.

3 M Bruce, *New Technology and the Future of Tourism*. Tourism Management. June 1987.

4 M Bruce, *Information Technology: Changes in the Travel Trade*. Tourism Management. December 1983.

10 Privacy and security of videotex systems

1 H Maurer, *The Austrian approach to videotex*. Proceedings of the 7th European meeting on cybernetics & system research, pages 589–592. North Holland, 1984.

2 H Maurer and H Cheng, *Teleprograms—the right approach to videotext—if you do it right*. Proceedings of the IRE conference on telesoftware, pages 75–78. London, 1984.

3 H Maurer and I Sebestyen, *On some unusual applications of videotex*. Proceedings of Videotex 82, New York, pages 199–210. Online, 1982.

4   H Maurer and I Sebestyen, *Public-key cryptosystems, telesoftware and other novel applications of videotex*. Proceedings of Viewdata 82, London, pages 145–158. Online, 1982.

5   M Shaine, *Microcomputer publishing*, Proceedings of the IRE conference on telesoftware, pages 59–69. London, 1982.

6   M Shaine, *Going for the microcomputer market with commercial software*. Proceedings of Viewdata 82, London, pages 119–134. Online, 1982.

7   H Maurer, *Wie nah ist 1984?* Politikum 18, pages 8–11. 1984.

8   H Maurer, N Rozsenich and I Sebestyen, *Videotext without Big Brother*. Electronic Publishing Review number 4, pages 201–214. 1984.

22   Broadband communications, a global view

1   *New Media Markets*, Volume 6 Number 4. Financial Times, March 1984.

25   On-demand interactive video

1   W K Ritchie, *The British Telecom Switched-Star Cable-TV Network*, British Telecom Technology Journal, Volume 2 Number 4, September 1984.

2   C Bayard-White, *Interactive Video Case Studies and Directory*. Council for Educational Technology and National Interactive Video Centre, December 1985.

3   A Luther, *New Integrated Video and Graphics Technology: Digital Video Interactive*. Optical Information Systems, Volume 7 Number 6. November–December 1987.

28   VSATs and small dishes: business or pleasure

1   W E Strich, *VSAT Networks: the US Experience*. European Statellite Communications Conference. Online, December 1987.

30   Towards the intelligent satellite

1   L Blonstein, *Private Satellite Networks in the USA: Lessons for Europe*. Satellite Communnications & Broadcasting conference. Online, December 1986.

2   J L Rose and H Orenstein, *The use of Small Customer Premises Terminals to Establish Multipoint Satellite Communications Networks*. IEE Colloquium on Small Terminal Satellite Communication Systems. IEE, November 1986.

3   S J Amy, N Barton and V Cheong, *US Bypass Communications Marketplace*. Scicon Ltd, December 1986. Unpublished.

4   J Collins, *Comsat's Experience of and Involvement in the Market for VSATs in the USA and Europe*. Eurosatellite 87. IBC Legal Studies & Services Ltd, London, May 1987.

5   *Contel ASC Acquisitions Spur Market Realignments*. Satellite Communications (USA), Volume 11 Number 9, page 10. September 1987.

6   *Private Networks 1987*. Satellite Communications (USA), Volume 11 Number 5, pages 23–26. May 1987.

7   E Youssefzadeh and G Baley, *Are VSATs a viable bypass alternative for two-way data traffic? Pros and Cons*. Network World. 19 January 1987.

8   D Friedman, *How to Avoid Pitfalls and Pratfalls when Buying Your Own VSAT Satellite Network and Service*. Communications News, Volume 24 Number 3, pages 30–35. March 1987.

9   *Intelnet Service Applications, Mercury Communications—The UK VSAT Pioneer*. Intelsat Business Services seminar, Munich. November 1986.

10 *SatStar, British Telecom's Interactive Satellite Data Network*. British Telecom International. London.

11 D J Shorrock, *Are VSAT Data Networks Viable?—Technical, economic and regulatory issues*. Eurosatellite 1987. IBC Legal Studies & Services Ltd, London, May 1987.

12 *Towards a Dynamic European Economy. Green Paper on the development of the Common Market for telecommunications services and equipment*. Communication from the Commission to the Council, Commission of the European Communities, Telecommunications, Information Industries and Innovation. DG XIII. 9 June 1987.

13 *Eutelsat agrees to recognise Astra television satellite*. Financial Times, Number 30,350, page 2. 30 September 1987.

Appendix C

Guide to Acronyms

The use of acronyms in the Information Technology industry is a widespread reality: the papers presented in this book are no exception. In most cases the authors explain the acronyms they use the first time they use them, but not subsequently. The following is a list of the more widely applicable acronyms used.

ABTC	Adaptive Block Truncation Coding
ACD	Automatic Call Distributor
ADRMPS	Auto-dialed Recorded Message Players
ALEX	Public Videotex Service in Canada
ANTIOPE	l'Acquisition Numérique et Télévisualisation d'Images Organisée en Pages d'Ecriture
ATD	Asynchronous Time Division
ATDM	Asynchronous Time Division Multiplexing
ATM	Automated Teller Machine
BBC	British Broadcasting Corporation
BPO	British Post Office
BPSK	Binary Phase Shift Keyed
BT	British Telecom
CAD/CAM	Computer Aided Design/Computer Aided Manufacturing
CAPE CAPTAIN	Editing Unit
CAPF CAPTAIN	Information Processing Unit
CAPTAIN	The Character and Pattern Telephone Access Information Network
CATV	Community Antenna Television
CCC	Chaos Computer Club
CCITT	Comité Consultatif International Télégraphique et Téléphonique
CCS	Common Channel Signalling
CCTV	Croydon Cable Television
CD	Compact Disc
CD-A	Compact Disc Audio
CD-DVI	Compact Disc Video Interactive
CD-E	Compact Disc Erasable
CD-I	Compact Disc Interactive
CD-ROM	Compact Disc Read Only Memory
CD-RTOS	Compact Disc Real Time Operating System
CD-V	Compact Disc Video
CDMA	Code Division Multiple Access

CEC	Commission of the European Communities
CEPT	Conference Européenne de l'Administration des Postes et des Télécommunications
CLV	Constant Linear Velocity
CMOS	Complimentary Metal Oxide Semiconductor
CNET	Centre Nationale d'Etudes de Télécommunications
CPU	Central Processing Unit
CRB	Customer Reconfigurable Bandwidth
CUG	Closed User Group
DARPA	Defence Advanced Research Project Agency
DAT	Digital Audio Tape
DATAPAC	National Packet Switched Network of Denmark
DBS	Direct Broadcast Satellite
DDX-PS	Digital Data Exchange Packet Switch
DF	Direct Access Information Centre
DFB	Distributed Feedback
DGT	Direction Générale des Télécommunications
DLS	Digital Local Switch
DM	Deutsche Mark
DRAW	Direct-Read-After-Write
DTH	Direct To Home
DTI	Department of Trade & Industry (UK)
DTMF	Dual Tone Modulation Frequency
DV-I	Digital Video Interactive
DVCP	Digital Videotex Communication Processing Unit
ECL	Emitter Coupled Logic
ECS	European Communications Satellite
EDI	Electronic Data Interchange
ESA	European Space Agency
ET	Simple Editing Input Terminal
FDMA	Frequency Division Multiple Access
FF	French Franc
GAP	EEC Analysis & Forecasting Group
GDH	Global Digital Highway
GS	Gateway Switch
HDTV	High Definition Television
HSG	High Sierra Group
IBCN	Integrated Broadband Communications Network
IBS	International Business Services (IBM Global Teleprocessing Network)
IF	Indirect Access Information Centre
INC	Information Input Centre
IP	Information Provider
ISDN	Integrated Services Digital Network
IT	Information Input Terminal
IT	Information Technology
ITU	International Telecommunications Union
IVDT	Interactive Voice/Data Terminals
IVIS	Interactive Video in Schools

LAMI	French Professional Transportation Service
LAN	Local Area Network
LCD	Liquid Crystal Display
LNC	Low Noise Converter
M3VDS	Millimetric Microwave Video Distribution System
MAN	Metropolitan Area Network
MAP	Manufacturing Automation Protocol
MATV	Master Antenna Television
MMDS	Multipoint Microwave Distribution System
MTVDS	Microwave Television Distribution System
MVDS	Microwave Video Distribution System
NASA	National Aeronautical & Space Administration
NTT	Nippon Telephone and Telegraph
NTV	National TeleVoice
OBP	On Board Processing
PABX	Public Automated Branch Exchange
PBX	Private Branch Exchange
PC	Personal Computer
PIN	Personal Identification Number
PNG	Peacock National Grid
PRE	Picture Response Equipment
PSTN	Public Switched Telecommunications Network
PTT	Post, Telegraph and Telephone Administration
QPSK	Quadrature Phase Shift Keying
RACE	Research in Advanced Communications Technologies in Europe
SCPC	Single Channel Per Carrier
SMATV	Satellite Master Antenna Television
START	West German Information & Reservation System
TAN	Transaction Number
TBB	Transnational Broadband Backbone
TDMA	Time Division Multiple Access
TOPS	Thompson Holiday's Videotex Booking System
TRANSPAC	National Packet Switched Network of France
TSB	Trustee Savings Bank (UK)
TUI	Touristik Union International (West German)
VAN	Value Added Network
VANS	Value Added Network Service
VAR	Value Added Reseller
VCP	Videotex Communication Processing Unit
VCR	Video Cassette Recorder
VLSI	Very Large Scale Integration
VODIS	Voice Operated Database Inquiry Service
VRU	Voice Response Unit
VSAT	Very Small Aperture Terminal (Microterminal)
WAN	Wide Area Network
WORM	Write Once Read Many Times